PROSPECTOR'S
Field Book & Guide
1901

H. S. Osborn

reprinted by Lindsay Publications Inc

Prospector's
Field Book & Guide

by H. S. Osborn

Originally published by
Henry Carey Baird & Co
Philadelphia

Original copyright 1901
by Henry Carey Baird & Co

Reprinted by
Lindsay Publications Inc
Bradley IL 60915

ISBN 1-917914-57-0

5 6 7 8 9 0

1987 1999

THE PROSPECTOR'S

FIELD-BOOK AND GUIDE

IN THE

SEARCH FOR AND THE EASY DETERMINATION OF
ORES AND OTHER USEFUL MINERALS.

BY

Prof. H. S. OSBORN, LL.D.,

AUTHOR OF "THE METALLURGY OF IRON AND STEEL," "A PRACTICAL MANUAL
OF MINERALS, MINES, AND MINING."

ILLUSTRATED BY FIFTY-EIGHT ENGRAVINGS.

FIFTH EDITION, REVISED AND ENLARGED.

WARNING

Remember that the materials and methods described here are from another era. Workers were less safety conscious then, and some methods may be downright dangerous. Be careful! Use good solid judgement in your work, and think ahead. Lindsay Publications, Inc. has not tested these methods and materials and does not endorse them. Our job is merely to pass along to you information from another era. Safety is your responsibility.

Write for a catalog or other unusual books available from:

Lindsay Publications, Inc.
PO Box 12
Bradley, IL 60915-0012

PREFACE TO THE FIFTH EDITION.

THE gratifying success of the fourth edition of THE PROSPECTOR'S FIELD-BOOK AND GUIDE, unmistakably indicating the firm hold which it has on the confidence of Prospectors, has rendered necessary the preparation of this, the fifth edition. In doing this, the book has been carefully revised throughout, and where considered desirable, it has been enlarged. These revisions and amplifications add greatly, as it is believed, to the value and usefulness of the volume, and bring it fully up to date.

The work of revision has been undertaken by the same competent hand that so satisfactorily edited the second, third and fourth editions. As now presented to the public, it is felt to be a complete and thoroughly reliable guide and companion to the intelligent and enterprising searcher after ores and other useful minerals, including gems and gemstones. It has been provided with a thorough Table of Contents and an Index, rendering reference to any subject in it prompt and easy.

The publishers confidently look for a sale of this edition equal in its rapidity and extent to that of those which have preceded it.

H. C. B.

PHILADELPHIA, July 1, 1901.

PUBLISHER'S PREFACE TO THE SECOND EDITION.

THE death of Dr. Osborn, two years ago, renders it necessary that the Publisher should prepare the preface to this revised edition of THE PROSPECTOR'S FIELD-BOOK AND GUIDE.

The fact of a second edition of this book having been called for so soon after the publication of the large first edition, justifies the belief that it has supplied a public requirement. The task of revising the work has devolved upon thoroughly competent hands; and whilst it has been aimed, by the insertion of further information regarding the subjects treated in the original edition, to make it still more acceptable to those for whom it was prepared, a new chapter has also been added on Petroleum, Ozocerite, Asphalt and Peat, together with a Glossary of Terms used in prospecting, mining, mineralogy, geology, etc.

While the work of revision has been done with conscientious care, under the supervision of the Publisher, it can hardly be hoped that it has been so well done as if Dr. Osborn, with his profound knowledge of the subject treated, had been alive to direct it for himself, and in his own manner.

(vii)

Henry Stafford Osborn was born in Philadelphia, August 17, 1823, and died in New York City, February 2, 1894. He was graduàted at the University of Pennsylvania in 1841 ; went abroad in 1843 or 1844 ; studied at Bonn, Germany, and at the Polytechnic Institution of London. Before the civil war he held the chair of Natural Science at Roanoke College, Va., and in 1866 accepted a professorship at Lafayette College, Easton, Pa. Leaving Lafayette in 1870, he became, in 1871, Professor in Miami University at Oxford, Ohio. In 1865 he received from Lafayette College the degree of LL.D.

In 1869 he published " The Metallurgy of Iron and Steel ;" in 1888, " A Practical Manual of Minerals, Mines and Mining ;" in 1892, the first edition of THE PROSPECTOR'S FIELD-BOOK AND GUIDE, the success of all of which books has been pronounced.

Personally, Dr. Osborn was charming, full of information on a wide range of subjects, which he had studied thoroughly ; enthusiastic, amiable and just ; and the relations of his publisher with him during a quarter of a century, will ever be among the brightest and best recollections of that publisher's long career in business.

HENRY CAREY BAIRD.

PHILADELPHIA, January 15, 1896.

PREFACE TO THE FIRST EDITION.

In the following pages we have attempted to present such a view of the whole subject of prospecting for the useful minerals that any liberally educated reader may fully comprehend our meaning. We have therefore explained special terms where we have thought it convenient to use them, and where the technically educated student would not need an explanation.

It must be understood that the subjects of chemistry, mineralogy, and metallurgy are introduced only for their practical bearing upon the ores in hand, or those sought for, and not for theory, or the philosophy of the operation, much as such theory or philosophy would please and instruct. The prospector must, therefore, refer to larger works if he desire to be instructed in the principles governing the sciences, the teachings of which we have frequently made use of.

We would suggest to any one intending to use this volume for practical work, to become acquainted with the whole book before attempting to use any special part alone. The object and construction have made it necessary to treat some special topics without repeating principles and methods already given in some part of the work, but which bear some relation to the topic under immediate consideration.

The Table of Contents and Index have both been carefully prepared, and being very full, will make reference to any subject in the volume easy and satisfactory.

OXFORD, OHIO, Jan. 5, 1892.

CONTENTS.

CHAPTER I.

PREPARATORY INSTRUCTIONS.

(xi)

CHAPTER II.

THE BLOWPIPE AND ITS USES.

CHAPTER III.

CRYSTALLOGRAPHY.

CHAPTER IV.

SURVEYING.

CHAPTER V.

ANALYSES OF ORES.

CHAPTER VI.

SPECIAL MINERALOGY.

CHAPTER VII.

TELLURIUM, PLATINUM, SILVER.

CHAPTER VIII.

COPPER.

CHAPTER IX.

LEAD AND TIN.

CHAPTER X.

ZINC—IRON.

CHAPTER XI.

MERCURY, BISMUTH, NICKEL, COBALT AND CADMIUM.

CHAPTER XII.

ALUMINIUM, ANTIMONY, MANGANESE.

CHAPTER XIII.

VARIOUS USEFUL MINERALS.

CHAPTER XIV.

PETROLEUM, OZOCERITE, ASPHALT, PEAT.

CHAPTER XV.

GEMS AND PRECIOUS STONES.

APPENDIX.

WEIGHTS AND MEASURES, SPECIFIC GRAVITY, BORING, CHEMICAL ELEMENTS, GLOSSARY, ETC.

THE

PROSPECTOR'S FIELD-BOOK AND GUIDE.

CHAPTER I.

PREPARATORY INSTRUCTION.

It is well-known that much disappointment and loss accrue through lack of knowledge by prospectors, who, with all their enterprise and energy, are often ignorant, not only of the probable locality, mode of occurrence and widely differing appearance of the various valuable minerals, but also of the best means of locating and testing the ores when found. It is a well-established fact that the majority of the best mineral finds have been made by the purest accident, often by men who had no mining knowledge whatever, and that many valuable discoveries have been delayed, or, when made, abandoned as not payable from the same cause—ignorance of the rudiments of mineralogy and mining. Hence in preparation for skilled work, the prospector should have become thoroughly acquainted with the forms under which useful minerals and metals appear.

This should be his very *first study*. It may be called the study of TECHNICAL MINERALOGY.

(1)

Minerals are composed of chemical elements, which are substances which cannot be further separated. A table of the chemical elements, their symbols, equivalents and specific gravities, is given in the Appendix. When these elements unite together and form a compound, they always do so in fixed proportion and in definite weight. Therefore, in any pure mineral, whose composition is known, the amounts of the elements going to make up any given mass of it can be calculated by a rule of three sum.

For example, in galena (PbS) we have lead (Pb) $=207$ and sulphur (S)$=32$, total 239. Therefore, in 239 lbs. of pure galena we will find 207 lbs. of lead ($86\frac{1}{2}$ per cent.), and so on in proportion.

Thus any mineral that is pure enough to be weighed directly, or which can be concentrated pure and then weighed, can be estimated in this way, and the percentage content of the ore calculated.

The combination of two or more of these elements together gives rise to three classes of substances, namely, *acids, bases and salts.*

Oxides of non-metallic elements are acid.

Oxides of metallic elements are bases.

Where an acid and a base unite, one exactly neutralizing the other, a substance is produced having neither acid or basic tendency. It is known as a salt.

Most minerals are salts. There is only one common acid mineral, namely, quartz (S_2O_2), or the oxide of the non-metallic element silicon.

There are many minerals which are basic, such as hematite (Fe_2O_3) and magnetite (Fe_3O_4), the oxides of iron, and cuprite (CuO), the oxide of copper.

Among the many minerals which are salts are common salt or sodium chloride (NaCl); limestone or calcite $(CaCO_3)$, formed from the union of the oxide of calcium (metal) and carbonic acid gas; gypsum $(CaSO_4 2H_2O)$, formed by the union of the oxide of calcium (metal) and sulphuric acid; apatite, phosphate of lime $[Ca_3(P_2O_4)_2]$, formed by the same base as above uniting with phosphoric acid.

There are a great many minerals the acid member of which is silica, with one or more metallic oxides forming the basic member. These are known as *silicates*, and felspar, mica, hornblende, pyrosene, talc, serpentine, etc, are examples.

These facts are important to remember, because whole families of minerals and rocks are classified acid or basic according to the greater or lesser quantity of silica present in them.

Some metals are found native and in some degree of purity, as in the cases of gold, silver, copper, mercury, and platinum, and when so found are readily determined at once by any one who is at all acquainted with those metals as they occur in general use. But frequently native metals appear under such colors, and even forms, that the discoverer must possess more knowledge than any one usually possesses who has seen the metal in the arts only. Gold, as an illustration, is frequently found in various shades of yellow, in accordance with the amount of silver or copper it may contain, and yet to the practiced eye of a true mineralogist it never loses the true gold hue.

Iron pyrites, which is composed of sulphur and iron, and called "pyrite," mineralogically, has a color somewhat similar to that of gold, and so also has a mineral called "chalcopyrite," or copper pyrites, which contains copper, iron and sulphur. These, with others, vary in the yellow shade, and degrees of color, but by the practiced eye are instantly detected. Of course the brittleness of these minerals is unlike the softness of native gold, and this would instantly reveal the fact that they were not gold; but we are now speaking of the practiced eye alone, and therefore of the benefit of cultivating a knowledge by sight of minerals. The mode in which a mineral breaks when smartly struck with a hammer, or pressed with the point of a knife, is a character of importance. Many minerals can only be broken in certain directions, for instance, a crystal of calc spar can only be split parallel to the faces of a rhombohedron; many crystals break more readily in one direction than in others. Whenever a mineral breaks with a smooth, flat, even surface, it is said to exhibit *cleavage*. Cleavage always depends upon the crystalline form. But minerals often break in irregular directions, having no connection whatever with the crystalline form, and this kind of breaking is called *fracture*. The nature of the surface given by fracture is often a character of importance, especially in distinguishing the varieties of a mineral species. Thus quartz and many mineral species show a shell-like fracture-surface which is called *conchoidal,* or if less distinct, *small conchoidal* or *sub-conchoidal.* More commonly the fracture is simply said to be *uneven,* when the surface is rough and irregular. Occasionally it is *hackly,* like a piece of fractured iron. *Earthy* and *splintery* are other terms sometimes used and readily understood.

Streak. The color and appearance of the line or furrow on the surface of a mineral, when scratched or rubbed, is called the streak, which is best obtained by means of a hard-tempered knife or a file. The color of a mineral and its streak may correspond, or the mineral and its streak may possess different colors, or the mineral may be colored while its streak is colorless. For instance, cinnabar has both a red color and a red streak; specular iron has a black color, but a red streak; sapphire has a blue color, but a white colorless streak. The streak of most minerals is dull and pulverulent, but a few exhibit a shining streak like that formed on scratching a piece of lead or copper. This kind of streak is distinguished by the name of *metallic.* In judging the streak of a mineral, much-weathered pieces should be rejected.

Hardness is another character of great importance in distinguishing minerals; it is the quality of resisting abrasion. The diamond is the hardest substance known, as it will scratch all others. Talc is one of the softest minerals. Other minerals possess intermediate degrees of hardness. To express how hard any mineral is, it becomes necessary to compare it with some known standard. Ten standards of different degrees have been chosen, and are given in order in the following scale:

1. *Talc*, easily scratched by the finger-nail.

2. *Gypsum*, does not easily yield to the finger-nail, nor will it scratch a copper-coin.

3. *Calcite*, scratches a copper coin, but is also scratched by a copper coin.

4. *Fluorite*, is not scratched by a copper coin, and does not scratch glass.

5. *Apatite*, scratches glass with difficulty; is readily scratched by a knife.

6. *Feldspar*, scratches glass with ease; is difficult to scratch by a knife.

7. *Quartz*, cannot be scratched by a knife, and readily scratches glass.

8. *Topaz,*
9. *Corundum,* } harder than flint or quartz,

10. *Diamond*, scratches any substance.

If on drawing a knife across a mineral it is impressed as easily as calcite, its hardness is said to be 3. If a mineral scratches quartz, but is itself scratched by topaz, its hardness is between 7 and 8.

In trying the hardness of a mineral, a sound portion of the mineral should be chosen and a sharp angle used in trying to scratch. A streak of dust on scratching one mineral with another may come from the waste of either, and it cannot be determined which is the softer until after wiping off the dust and examining with a lens.

By the test of hardness, clear distinctions may be drawn between minerals which resemble each other. Iron pyrites and copper pyrites, for instance, are similar in appearance, but copper pyrites can easily be scratched with a knife, while iron pyrites is nearly as hard as quartz and the knife makes no impression upon it.

Flexibility and elasticity. Some minerals can be readily bent without breaking, for instance, talc, mica, chlorite, molybdenite, native silver, etc. Minerals which after being bent can resume their former shape like a steel spring, are called elastic, for instance, mica and elaterite. A remarkable instance of flexibility, even combined with elasticity, amongst the rocks, is that of a micaceous sandstone called itacolumite, which in Brazil is the matrix of the diamond.

Smell. A few minerals only, like bitumen, have a strong smell which is readily recognized, but specimens generally require to be struck with a hammer, rubbed, or breathed upon before any smell can be observed. Some black limestones have a bituminous odor, while some have a sulphurous, and others a fœtid, smell. Hydraulic limestone has a smell of clay which can be detected when the mineral is breathed on. Some minerals containing much arsenic, for instance mispickel, smell of garlic when struck with a hammer.

Taste. Only soluble minerals have any taste, and this can only be described by comparison with well-known substances, for instance *acid*, vitriol ; *pungent,* sal ammoniac; *salt*, rock salt; *cooling*, nitre; *astringent*, alum; *metallic astringent*, sulphate of copper; *bitter*, sulphate of magnesia; *sweet*, borax.

Malleability. Malleable substances can be hammered out without breaking, and it is on this quality that the value of certain metals in the arts depends, for instance, copper, silver, gold, iron, etc.

A few minerals are malleable, and at the same time sectile, *i. e.*, they can be cut with a knife, for instance, silver glance, horn silver and ozokerite.

Mineral caoutchouc (elaterite) is sectile, but like india rubber, can only be shaped when hot. The elasticity of elaterite is so characteristic that the mineral will be readily recognized.

Ductility, or the capability of being drawn into wire, is a property which is confined exclusively to certain metals. It is possessed in the highest degree by gold, which can be drawn into the finest wire, or rolled into leaves of such fineness that 30,000 of them are not thicker than an eighth of an inch.

Lustre. The term lustre is employed to describe with certain adjectives, the brilliancy or gloss of any substance. In describing the lustre well-known substances are taken as the types, and such terms as *adamantine lustre*—diamond-like—and *vitreous lustre*—glassy—are used. The lustre of a mineral is quite independent of its color. When minerals do not possess any lustre at all they are described as "dull." The kinds of lustre distinguished are as follows:

Metallic: The lustre of a metallic surface as of steel, lead, tin, copper, gold, etc.

Vitreous, or *glassy* lustre : That of a piece of broken glass. This is the lustre of most quartz and of a large part of non-metallic minerals.

Adamantine: This is the lustre of the diamond. It is the brilliant, almost oily, lustre shown by some very hard materials, as diamond, corundum, etc. When sub-metallic it is termed *metallic adamantine*, as seen in some varities of white lead ore or cerussite.

Resinous or *waxy:* The lustre of a piece of rosin, as that of zinc blende, some varieties opal, etc. Near this, but quite distinct, is the *greasy lustre*, shown by some specimens of milky quartz.

Pearly or the lustre of mother-of-pearl. This is common where a mineral has very perfect cleavage. Examples : Talc, native magnesia, stilbite, etc.

Silky, like silk. This is the result of fibrous structure, as the variety of calcite (or of gypsum) called satin spar, also of most asbestus.

Specific gravity. Prospectors soon acquire some proficiency in testing the weight of minerals by handling them. A lump of pyrite, for instance, can readily be distinguished from gold by its weight, since a mass of gold of the same size would weigh at least three times as much, and a little practice with well-known substances will enable the prospector to class most minerals within certain broad limits by weighing them in the hand.

The specific gravity of a mineral is its weight compared with water at a standard temperature and pressure, which is taken as the standard, and de-

scribed as having a specific gravity of 1; conse-
quently, to determine that of a mineral, it is necessary to find the weight of a piece of the mineral and
that of a corresponding bulk of water, and to divide
the first by the last. This can be done with great
accuracy in the laboratory, where delicate balances
are available, but is not applicable in the field, when
the most that can be undertaken is to class minerals
roughly within certain broad limits, and indeed,
this is generally sufficient for the prospector. Some
rules for finding weights by specific gravity are
given in the Appendix.

What has previously been said of color may also
be said of weight and form. A lump of pyrite in
the hands of a skillful mineralogist would be distinguished from gold by its weight, since as above
mentioned, a mass of gold of the same size would
weigh at least three times as much. Three crystalline pieces, the one of barite, the other two of lime
carbonate and of quartz, may to the unskillful eye
appear equally transparent; but the form of the
first is tabular, that of the latter two is in six-sided
crystals, but the lime carbonate crystals terminate
in three sides, while the quartz always (like the
sides) in six.

Besides a knowledge of the forms under which the
minerals we seek present themselves, it is also necessary to learn the characteristics of some of the rocks
which are generally associated with those minerals.
The object of this knowledge is to serve in directing
us to those regions where we may with greater prob-

ability discover the minerals we seek. It also serves to warn us out of a region where we should not expect to find what we desire.

To illustrate, we may not expect to find iron ores of a certain kind, brown hematites for instance, in a granitic country. On the other hand, we may find the magnetic ores in such a region, and it is useless to explore a granitic region for black band iron ore, although it may be the proper region to discover red hematite.

It is, therefore, important that the prospector should be able to distinguish many of the geologic rocks to help in guiding or in checking him, in his explorations.

A general knowledge, therefore, of the manner in which the geologic rocks are "laid down," their order, or succession, in the earth, is important, and the distinction between sedimentary and that which has been, and is usually called "igneous rock," but more properly "azoic rock," that is, rock which does not exhibit any remains of fossil or organic life. For often the only signs by which we can, with any degree of certainty, determine what is the name of the sedimentary rock is by finding the remains of former life, that is, the kind of fossil it contains. Prof. Dana says (The Amer. Journal of Science, Nov. and Dec., 1890) that it is settled that the kind of rock in itself considered is not a safe criterion of geological age.

If all the rocks in the world had been laid down in regularly horizontal sequence and had always remained in their own separate "horizons," as every rock of the same age is called, not only should we find them all parallel, one over the other, but we might readily determine to some extent what were the exact order and distance of any one horizon, or geological age. But, although there is a general order, the same in all parts of the world, there have

FIG. 1.

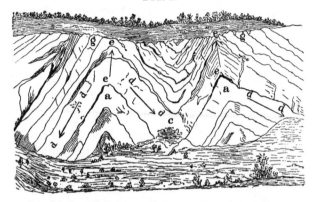

SECTION SHOWING CONTORTED STRATA DUE TO LATERAL PRESSURE, *aa*, "anti-clinal axis;" *c*, the "synclinal axis." The direction of the arrows, *ee*, *ee*, is that of "the strike." That of the arrows, *dd*, is that of "the dip" of the strata, always measured from the horizon; *gg*, are the out-crops.

been upheavals and sinkings, dislocations and ero-sions, during the ages, so that it is necessary that the prospector should become acquainted with the various changes probable in the order and forms of the vast rocks which carry the minerals for which he is seeking.

Some of these movements of the earth's crust are represented in Fig. 1.

PRACTICAL GEOLOGY.

We repeat that it is of considerable importance that the prospector should have at least some general knowledge of those geological horizons with which his work is specially associated. As we have intimated, useful minerals do not always confine themselves to one horizon; but there are certain ranges of rock which indicate their vicinity. There are also limits which are never overpassed by some useful minerals, and experience has shown that some horizons are always sterile in ores, and it is therefore useless ever to expect to find them in paying quantities, in certain rocks or beyond them in certain directions.

Gold often occurs where it will not pay to open and work the strata, so also with lead and copper. It is well to learn the relations of such barren regions, or horizons, as the strata are called.

In the following table we have given chief place to these horizons which have been found in our own country to abound in the useful minerals, and we advise the possession of small specimens of the principal rocks mentioned and the special examination of the specimens under a good lens, so as to become thoroughly acquainted with their appearance and their minute parts of composition.

All rocks may be classified as—

1. Igneous.
2. Metamorphic.
3. Aqueous.

Speaking geologically, not only the hard consoli-

dated, massive and stony substances are called
" rocks," but any natural deposits of stony material
such as sand, earth, or clay, when in natural beds,
are geological rocks. Very few of the rocks of this
earth, at any rate so far as examined, are in their
original and primal condition. Even the granites
and volcanic rocks are composed of other and more
ancient material disintegrated, ground up, or worn
down, settled, buried, and compressed by ages of
enormous pressure, or consolidated by cementation.
Some have been "laid down" under water, having
been disintegrated into dust, carried by the winds of
ages out over the oceans and seas, and settled down
into the form of the present rocks, which afterward
have been lifted up into mountains and plains above
the seas. But by the transporting power of rivers
or currents in ancient oceans, and because of unequal unheaval of some regions where subterranean
forces were greater than at distant places, very large
differences in the nature of the deposit have occurred,
even in limited regions. These special and limited
forces will account for the fact that although, taking
the geological horizons throughout the world, there
is a general sameness, differences do occur, and
important members of the order of succession are
omitted in some regions, and exceptions to general
rules occur.

We give, therefore, in the table following, those
universally accepted relations of certain rocks, one
to another, in the great geologic arrangement of the
world, omitting some of the subsidiary, limited, and
unimportant horizons.

1. IGNEOUS ROCKS are such as have been subjected to sufficient heat to melt the ingredi-
   ents. Of these rocks—

   Volcanic rocks are those which have been cooled
   near or at the surface, as lava, etc.

   Trachyte: A grayish rock of rough fracture ; the
   same specific gravity as quartz, but mainly
   constituted of grains of glassy feldspar. It
   is essentially a unisilicate of alumina, with
   10 to 15 per cent. potash, a little soda and
   lime ; differs from quartz in that it fuses
   before the blow-pipe, while quartz remains
   unfused except when soda is used.

   Basalt: Blackish or dark brown. *Traps, greenstone, dolerite, amydolite ;* these latter four are
   only modifications, being all unisilicates with
   smaller amounts of potash than in trachyte,
   a little more soda and lime, and some traces
   of iron and magnesia, varying in color and
   form.

   Obsidian is a glass, something like bottle glass,
   of a dark shade, and translucent.

All these are compact in texture except where
some holes have been worn in by steam or gases.
They are frequently found penetrating several strata,
having been forced up in columns almost vertically,
and sometimes spreading out horizontally for many
miles between the strata or on the surface, and are
called volcanic dykes, or intrusive rocks or lava.
These and such-like are igneous rocks.

It is not certain that granite rocks are of igneous

origin, but they seem to belong to the metamorphic series.

2. METAMORPHIC; these are of igneous, subsequently to the time when they were of aqueous origin, and have undergone a change through pressure and heat, and, perhaps, in connection with steam or water. Of this class are the following:

GNEISS, having a composition of small pieces of feldspar, mica, and quartz, like some granites, but laminated or foliated in form, and not equally solid, homogeneous, and continuous throughout its structure as granite is.

MICA SCHIST. This term is given to those laminated rocks composed of mica and quartz in small particles, easily broken up, but more easily broken into tabular or leaf-like pieces, because the mica has been deposited in planes allowing of cleavage.

3. THE AQUEOUS ROCKS are simple water rocks—that is, rocks composed of sediments from the dust or ground-up remains of other rocks. The presence of such sediments is due to the transporting power of rivers, floods, or currents, and also of winds and storms and other agencies, carrying the dust to the ocean waters where it was arrested and became a sediment.

In sandstone (Fig. 2), the grains of sand are rounded, having no sharp edges as in granite.

	GENERAL DIVISIONS.		SUBDIVISIONS.
TERTIARY OR CENOZOIC.	RECENT, PLEISTOCENE, OR QUARTERNARY.		All its shells and bones are of existing species.
	PLIOCENE.		About 50 per cent. of existing species of shells.
	MIOCENE.		Contains 80 per cent. of extinct species.
	EOCENE.		Contains fresh water and marine strata, animals all extinct.
SECONDARY OR MESOZOIC.	CRETACEOUS.		Upper. Middle. Lower.
			Whealden.
	JURASSIC.		Portland Stone. Oxford Group. Stonesfield Slate.
		Lias	Limestone in horizontal strata.
	TRIASSIC.		Keuper. Muschelkalk. Bunter-sandstone.
PRIMARY OR PALEOZIC.	PERMIAN.		Dark red sandstone. Magnesian limestone. Conglomerates, Breccias Marls in all three.
	CARBONIFEROUS.		Seams of Anthracite and bituminous coals of varying thickness. Millstone grit. Subcarboniferous.
	DEVONIAN.		Catskill Period. Chemung Period. Hamilton Period. Corniferous Period.
	SILURIAN.	Upper	Oriskany Sandstone. Lower Helderberg Period. Salina Period. Niagara Period.
		Lower	Trenton Period. Canadian Period. Potsdam Sandstone.
			Cambrian. Laurentian. ARCHÆAN.

STRATIFIED ROCKS.

CHARACTERISTICS.

Tertiary rocks yield brick and other clays, gypsum, sand, phosphate of lime deposits such as are in Florida, South Carolina, and elsewhere. GOLD in the drift and alluvial, also PLATINUM (Iridium, *see text*), and TIN.

Coal fields (brown or lignite) of this period, occur in India, Indian Archipelago, Japan, New Zealand, Vancouver's Island, and in Europe; also in California, Washington, Oregon, Colorado, etc. The true coal (anthracite and bituminous) belongs to the Carboniferous only.

A very hard lignite exists at Gay Head, Martha's Vineyard, in this formation.

Upper Chalk with flints, but the Lower ⎫ The whole formation contains sea-shells, sponges, Chalk without flints. ⎭ sea-urchins, etc.

Contains Greensand in England and in New Jersey, used as a marl and fertilizer. There is a supposed Cretaceous lignite in Alaska, Colorado, California, Utah, etc.

Consists of sand, clay, or marl, the sand used in glass making.

Some English coal is found in the Oolite. Kimmeridge clay is found in upper Oolite; the fine Bavarian lithographic stone in the middle Oolite.

Conspicuous for the number of ammonites and nautilus shells. Furnishes building and paving stone.

Called by the Germans TRIAS.
Connecticut river sandstone with footprints.
Red clays, marls, shales and sandstones. The New Red Sandstone of England.
In Europe great salt,beds.

Mostly sandstones and marlytes, some impure magnesian limestone and gypsum. Thin seams of coal, unworkable. With exception of BROWN HEMATITE iron ore and the metals mentioned above, all the other metals are found in the formations below.

The black band iron ore. Limestone from the same mines with the coal in Great Britain, but not so frequently in America. Anthracite, cannel, and bituminous coal in seams in limestone, sandstone, and shales, forming the "The Coal Measures."

Affords PETROLEUM in Pennsylvania, Ohio, and elsewhere, and salines in Michigan. It is the MOUNTAIN LIMESTONE of England. Largely of corals.

Includes the OLD RED SANDSTONE OF ENGLAND.
Hamilton black shales produce oil; the Hamilton beds afford excellent flagging stone.
Corniferous called also Upper Helderberg group.

Salina Period supplies the salt waters of Salina and Syracuse, N. Y.

The LEAD MINES of Iowa and Wisconsin are in the Magnesian Limestone of the Canadian Period.

(Between pages 16 and 17.)

Where the sedimentary material was exceedingly dust-like, it sometimes is laid down as fine mud and frequently in lamina, as in shale (Fig. 3).

FIG. 2.

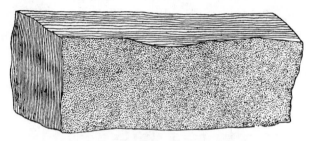

Sandstone.

GRANITE is a term descriptive of rocks generally composed of quartz, feldspar and mica, in grains (hence the name) of a crystalline form. But the

FIG. 3.

Shale.

granites are not all alike in the amount of either of the above-mentioned minerals, nor are they alike in color. Some granites contain no mica, as in *graphic* granite, only quartz and feldspar, and the quartz in

2

the feldspar resembling written characters. Others contain hornblende as well as mica, or in the place of mica; the hornblende being in dark or black crystalline specks, pieces, or crystals, and consisting essentially of silica, magnesia, lime, and iron. This granite is called *syenite* granite. Where the feldspar is in distinct crystals in compact base, and sometimes lighter than the base, which is frequently reddish, purple, or dark green, it is a *porphyritic* granite. The granites are sometimes whit-

FIG. 4.

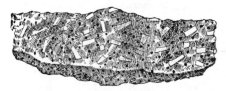

Granite with black mica and feldspar crystals, with quartz as chief base.

ish, grayish, or flesh-red. They are considered as metamorphic and not igneous (Dana), although some authors still consider them to be igneous. They always present a crystalline grain in varying degrees of fineness and prominence. One form is given in Fig. 4, from a specimen in the author's possession.

This specimen contains two kinds of mica, one black, *biotite*, the other white, of silvery appearance, *muscovite*. The biotite presents in spots the appearance of hornblende, and only the pen-knife point shows the scaly lamination of mica under the lens. It also contains crystalline forms of potash feldspar (*orthoclase*), distinguishable from the quartz by their side only, by the lamellar fracture of its edges, and its peculiar vitreous glimmer, for practically the hardness appears the same, although feldspar is (6.6 and quartz 7) slightly softer. It would be well for the prospector to gather many forms of granite and examine them under the lens until he becomes thoroughly used to the variations.

The first indications of a deposit possessing economic value are, as a rule, to be met with among the materials forming the beds of streams, and wherever water courses have seamed and furrowed the rocks. Metalliferous deposits should be looked for in hilly districts as a general rule, though alluvial accumulations may be found in comparative flat country. A close study of natural phenomena will often help in the discovery of mineral wealth. Thus the form and color of the surface; stained patches; springs of water whether sweet or mineralized; scum floating on water (petroleum, etc.); accumulations of earth brought to the surface by burrowing animals; changes in vegetation; behavior of the magnetic needle. These, however, only serve to indicate existence without reference to quantity or quality.

The valuable minerals and metal-bearing deposits of the earth occur as

Lodes. By a lode or vein is generally meant a fissure in the rocky crust of the earth which is filled with mineral matter. In Australia a vein is called a *reef* and in California a *ledge.*

Beds and layers. The most common of bedded deposits are those of coal. Many kinds of iron ore are found in beds, also some copper ores in shale, silver and lead ore in sandstone, etc. Beds and layers are also known as *strata, measures, sills, mines, bassets, delfs, girdles.*

Irregular deposits, such as *pockets,* etc., which lie sometimes in various formations. Contact deposits, net-work of veins, and where mineral is diffused through rocks, or in small cracks.

Surface deposits. By surface deposits are understood the beds of alluvium which more or less cover the face of every country. These beds have been chiefly created by various mechanical agents, which, after having degraded the higher rocks, carry the material which has thus been formed down to lower levels. By this process of degradation most mineral deposits are so comminuted that by their exposure to the atmosphere they are decomposed and destroyed. However, substances like cassiterite, platinum, gold, etc., not being so readily subject to decomposition, have, in consequence, been more or less preserved and buried among these superficial deposits. In observing deposits of this kind notice has to be taken of their general situation, area, thickness and richness. Often several beds may be ranged one above the other, in which case their relative values have to be determined. In tracing any particular deposit, as, for example, whilst ascending a valley, if the particles of ore increase in size and number, the prospector may expect that he is approaching their common origin. Another indication that he is near this point of origin will be that he shall find the mineral less worn.

Comprehensively speaking, all metals are found in the oldest rocks only, and the latter form the backbone, so to speak, of the main ranges of metalliferous countries. Therefore, the prospector in making his road towards the mountains will have to select a spot for starting actual operations. For this purpose a locality should be chosen, where the rocks are neither too hard nor too soft, nor should they be of too uniform a character. The country most deeply indented with gullies, cañons and gulches running parallel to one another offers the best chances of success. The region near the sources of the main rivers is generally the richest in metals and always the most easily prospected, requiring less labor and time in its examination, the loose debris and wash being of much lesser depth on account of the greater fall in the river and creek beds than at other portions of their courses.

Auriferous lodes are most likely to be met with near the headwaters of river systems, and very frequently the alluvial gold begins at or near the locality where a number of auriferous lodes exist. This is a very common occurrence, and may be in the great majority of cases relied upon.

When a river forks at its head into two or more branches, it is strange to say, the source of the gold will nearly always be found in the right-hand branch, geographically speaking. It may be mentioned that in determining the right and left-hand branches or banks of a river or stream, you are supposed to stand at the head of the river or stream looking towards its mouth or outlet. Amongst miners this is very often reversed, and quite a number of branches are named left-hand which, properly speaking, ought to be right-hand branches.

This right-hand theory is an old mining superstition for which science has offered no explanation, but the almost unfailing applicability of the theory is fully established by practical experience. Speaking of mining superstitions, it may be added that the spots upon which the sun shines before noon are held by miners to be richest in metal. Every old gold miner will pin his faith to this theory. What makes these observed facts—for they really amount to that—all the more remarkable is, that they may be applied with an equal degree of liability to the Northern and to the Southern hemispheres, which makes these superstitions appear in a paradoxial light. However, they have survived the test of hundreds of years in Cornwall and on the Continent of Europe, and have been confirmed by further observations in California and Australia. The latter instance, i. e., the spots upon which the sun shines before noon, may find an explanation in the fact that landslides and elevations of rock of all kinds are of more frequent occurrence upon the sunny than upon the shady side of valleys, the greater amount of disintegration of the rocks leading to a greater accumulation of the metals. However this may be, the theory forms one of the golden rules of the prospector.

The color of the rocks also serves as a guide to the prospector. Rocks of a pinkish-reddish color alternating with rocks of a deep bluish tint streaked with drab are generally very favorable to metallic deposits. Another good indication is when the faces of the precipices are covered with a black ooze caused by manganese, the presence of which always indicates a mineralized district. These are simply general indications.

Although color is always a good guide to the location of metallic deposits, it is of special service to the prospector in unexplored districts. Thus copper is indicated by greenish, bluish, or reddish stains upon the rocks in the neighborhood of the lode; tin and manganese, by dull black tints, manganese shows itself also in pinkish streaks. Gold, being always accompanied with iron, manifests its presence in red, yellow, or brown shades; lead and silver reveal grey or bluish-grey tinges; blende dyes the rocks yellowish-brown; and iron disports itself in all the hues of red, yellow-brown, and even dun-black.

The *wash of rivers and creeks,* and even more so that deposited upon terraces (if any) flanking the streams, must claim the close attention of the prospector. By wash is meant the diluvial drift in which gold or tin—the only metals mined in diluvial deposits—is found. The colors in connection with the different metals mentioned above, apply also to stones and the wash generally, though in a modified degree. Stones streaked with pinkish lines, and lines indicating manganese, are always found in wash conveying gold. Green stones, which are universally found in the wash, are always a good indication of gold if they are of a bright sea-green or even pea-green, but they must be smooth, hard, well-polished and very heavy. In many districts such stones are considered the "pilot stones" to gold. Quartz stones must be always present in goodly numbers in every gold-bearing wash, and if they are in a decaying state, they are all the better as a favorable indication.

Indicative plants. From very early times it has been noticed that the soil overlying mineral veins is favored by special vegetation, and though the occurrence of such vegetation cannot be taken as an infallible indication of the existence of such veins, it will be interesting to record the results of past observations, so that they may serve for a guidance to further observation in future.

Lead. The lead plant (*Amorpha canescens*) is said by prospectors in Michigan, Wisconsin and Illinois, to be most abundant in soils overlying the irregular deposits of galena in limestones. It is a shrub one to three feet high, covered with a hoary down. The light blue flowers are borne on long spikes, and the leaves are arranged in close pairs on stems, being almost devoid of foot stalks.

Gum trees, or trees with dead tops, as also sumac

and sassafras, are observed in Missouri to be abundant where " float " galena is found in the clays.

Iron. A vein of iron ore near Siegen, Germany, can be traced for nearly two miles by birch trees growing on the outcrop, while the remainder of the country is covered with oak and beech.

Limestone. The beech tree is almost invariably prevalent on limestone, and detached groups of beech trees have led to discoveries of unsuspected beds of limestone.

Phosphate. The phosphate miners in Estremadura, Spain, find that the *Convolvulus althæoides,* a creeping plant with bell-shaped flowers, is a most reliable guide to the scattered and hidden deposits of phosphorite occurring along the contact of the Silurian shales and Devonian dolomite.

Silver. In Montana experienced miners look for silver wherever the *Eriogonum ovalifolium* flourishes. This plant grows in low dense bushes, its small leaves coated with thick white down, and its rose-colored flowers being borne in clusters on long smooth stems.

Zinc. The " zinc violet," *Galmeiveilchen* or *Kelmesblume* (*Viola calaminaria*) of Rhenish Prussia, and neighboring parts of Belgium, is there considered an almost infallible guide to calamine deposits, though in other districts it grows where no zinc ore has been found. In the zinc districts its flowers are colored yellow, and zinc has been extracted from the plant. The same flower has been noticed at zinc mines in Utah.

In looking for indications where superficial deposits are known to occur, the prospector may be often guided, like the Tungusians in Northern Siberia, who search for gold by first looking at the general contour of the country, and observing those places where any obstacles, like a projecting range of hills would be likely to prevent material from being directly washed from higher to lower ground. Holes, sudden bends, or anything which would cause a diminution in the force of a current of water, are points at which it should be expected that heavy material like gold or platinum would be likely to collect. Although in Australia the most gold is generally found in *pot holes* and behind *hard bars*, it has often been found upon the shallow bends of ancient river courses. The lowest of a series of beds is generally the richest. In California the gold-bearing beds usually consist of gravels, which may be cemented to form a conglomerate, sands, bands of tuff, clay, fossil-wood, etc.

Magnetite occurs in alluvial deposits. Bog iron and manganese ore which have accumulated by precipitation in marshy places or in lakes usually contain too much impurity to be of commercial value. Stream tin occurs in gravels in much the same way as gold.

In examining a lode, the nature of the various minerals it contains and the proportions which these hold to each other should be observed. Sometimes it will be noticed that certain groups of minerals are often found together, the presence of one

being favorable to the existence of the other. At
other times the reverse will be remarked, the existence of one mineral being the sign of the absence of
another. The practical advantages to be derived
from a series of observations indicating such results
are too obvious to be overlooked.

The following table, showing the association of
ore in metalliferous veins, is given by Phillips and
Von Cotta:

Two Members.	Three Members.	Four or More Members.
Galena, blende.	Galena, blende, iron pyrites (silver ores).	Galena, blende, iron pyrites, quartz *and* spathic iron, diallogite, brown spar, calc spar *or* heavy spar.
Iron pyrites, chalcopyrites.	Iron pyrites, chalcopyrite, quartz (copper ores).	Iron pyrites, chalcopyrite, galena, blende; *and* spathic iron, diallogite brown spar, calc spar; *or* heavy spar.
Gold, quartz.	Gold, quartz, iron pyrites.	Gold, quartz, iron pyrites, galena, blende; *and* spathic iron, diallogite; brown spar, calc spar, *or* heavy spar.
Cobalt and nickel ores.	Cobalt and nickel ores, and iron pyrites.	Cobalt and nickel ores, iron pyrites; *and* galena, blende, quartz, spathic iron ore, diallogite, brown spar; calc spar; *or* heavy spar.
Tin ore, wolfram.	Tin ore, wolfram, quartz.	Tin ore, wolfram, quartz, mica, tourmaline, topaz, etc.
Gold, tellurium.	Gold, tellurium, tetrahedrite (various tellurium ores).	Gold, tellurium, tetrahedrite, quartz, *and* brown spar; *or* calc spar.
Cinnabar, tetrahedrite.	Cinnabar, tetrahedrite, pyrites (various ores of quicksilver).	Cinnabar, tellurium, tetrahedrite, pyrites, quartz; *and* spathic iron, diallogite, brown spar, calc spar; *or* heavy spar,
Magnetite, chlorite.	Magnetite, chlorite, garnet.	Magnetite, chlorite, garnet, pyroxene, hornblende, pyrites, etc.

CHAPTER II.

ALL chemical tests for minerals, whether with the blow-pipe or in the wet way, depend upon some chemical change, which is brought about, thus allowing the element, base or acid to be recognized. These changes consist either of the decomposition of the mineral, or the formation of fresh compounds. The following instances will sufficiently illustrate the character of these changes.

If the oxide of a metal, copper for instance, is mixed with carbonate of soda and fused on charcoal, the copper is reduced to a metallic state, the oxygen combining with the charcoal to form carbonic acid, which escapes as a gas, and any silica which is present decomposes the carbonate of soda to form a silicate of soda, which may be looked upon as a slag.

If a hydrous mineral is heated in a glass tube closed at one end, the water is given off, and condenses as drops in the cool part of the tube.

If an arsenical mineral is heated in a closed tube a crystalline deposit of arsenic is formed in the tube; but if it is heated in the air, white fumes of arsenious acid are evolved which smell like garlic.

If a drop of hydrochloric acid be placed on a car-

(28)

bonate, such as limestone, the presence of carbonic acid is recognized by the effervescence which takes place; the stronger acid having combined with the lime has liberated the carbonic acid in a gaseous form. In the case of very many mineral carbonates, the acid requires to be heated for this reaction.

A great deal can be learnt respecting a mineral by a few simple trials with the blow-pipe, and every prospector should learn to use it. The chief requirements are a plain brass blow-pipe about 7 to 10 inches long, a candle, a forceps or pliers, some platinum wire, a small pestle and mortar made of agate, a small sieve, a magnet, some small glass tubes, and some good firm charcoal free from cracks and openings.

The only reagents which will be absolutely necessary are borax, carbonate of soda and rarely microcosmic salt, nitrate of cobalt, and a little hydrochloric and sulphuric acid. A few others are occasionally necessary, but their use is limited. The carbonate of soda should be perfectly dry, not merely dry to the touch but quite free from water. Such carbonate of soda may be prepared from common washing soda by expelling the water it contains. Put the washing soda in a shallow, clean iron dish, and place it over a clear fire until a white dry powder is formed; avoid too strong a heat, otherwise the dry powder might fuse. A quarter of an ounce may be kept in a well-corked bottle or tube for use. Bicarbonate of soda may be used in-

stead without previous heating, or if the bicarbonate
be moderately heated it loses weight, and becomes
carbonate of soda, quite free from water, like the
above.

The borax is to be dried in the same way; a
quarter of an ounce will be enough. It is convenient to keep the platinum wire in the same tube.
Unless theses tubes are well corked, these chemicals
reabsorb moisture. For testing tin ore it is useful
to have a little cyanide of potassium kept in a bottle,
with the cork and rim well covered with melted
beeswax; it would otherwise liquefy by absorption
of moisture and become useless. It is a most dangerous poison, and the greatest caution must be observed in its use.

The blow-pipe should have a fine jet, or aperture,
wide enough to admit of a fine needle. The mode
of using it may be readily acquired by first breathing through the nostrils with the lips closed, then
puffing out the cheeks (as if rinsing the mouth with
water), still keeping the lips closed, and breathing
as before. The blow-pipe may at this point be
slipped between the lips, and it will be found that a
current of air escapes through it without any effort
on the part of the operator. Air flows through the
pipe owing to the tendency of the distended cheeks
to collapse; it must never be forced from the lungs.
After a little practice the strength of the current
may be increased. By breathing entirely through
the nostrils, keeping the lips closed, the blast may
be kept up for ten minutes or longer without ex-

haustion or inconvenience, except a slight fatigue of the lips in holding the blow-pipe. The beginner may practice blowing upon a piece of charcoal. The charcoal should, for convenience sake, be cut into slices of some six inches long by three-quarters to an inch wide and half inch thick. Place a piece of lead, or a pin-head, or fragment of pyrite (iron pyrites), near the end of the charcoal, and learn to blow the flame of a candle to a point upon the object. However awkward the blow-pipe may feel at first, practice will soon enable the learner to be expert. At first it may be necessary to gouge a small hole or recess in the coal with the point of your pen-knife, in order to prevent the specimen from being blown away. But after many trials such a command will be had over the blast that the hole may be made sufficiently deep by simply turning the point of the flame upon the coal and burning out a cavity.

Study the two colors of a sperm candle flame (Fig. 5). Notice that there is a yellow flame outside and nearer the top, and then within the flame there may be seen a bluish, probably a true blue flame. These flames act differently on the same substance. The outer *O F*, or yellow flame, is called the "*oxidizing flame*," the inner, the "*reducing flame*," *R F* or *I F*. By blowing properly, these two flames may be made to turn horizontally, or even downward, and then either the *O* flame or the *R* flame may be turned on the "assay" (as the object on the charcoal may be called). Get a piece of

iron ore as large as a pin-head and place it in a
little cavity on the charcoal, then cover it with a
quantity of soda carbonate as large as the assay.
Now turn the *R* flame down on the soda and ore,
and in a few seconds the ore will melt and be reduced to metallic iron, and your magnetized knife-
blade will pick it and the soda up.   In this experiment a piece of red or brown hematite, or a piece of

FIG. 5.

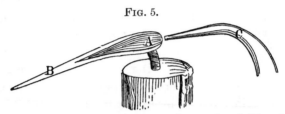

A, the blue or reducing flame ; *B*, the oxidizing flame ; *C*, the end of blow-pipe.

By placing the end of blow-pipe in the flame thus, the oxidizing flame, *A*, is
made more efficient.

pyrite (iron pyrites), should be used, as neither will
be attracted by the knife-blade *before* the ore is reduced to metallic iron.   The reason for this action
on the part of the ore is that the ore is metallic iron
combined with oxygen, and the *R* or blue flame calls
for more oxygen than it possesses, so that when it is
turned upon the hot oxide of iron it takes the

oxygen it calls for, from the ore and leaves the iron in a metallic state. But in the pyrite, which is iron and sulphur, the latter is partially driven off by either flame; and this process, on a larger scale, is called "*roasting*." The soda absorbs a part of the sulphur and part remains in the iron, but not so much but that the magnetized knife-blade will attract it. The last experiment is good for experimental practice, but not for illustrating the two properties of the flame.

The following is an excellent illustration and practice in showing the characteristic power of either flame. Get some platinum wire of the size of a large horse-hair. Wrap it around a match, leaving an end extending an inch and a half beyond the match end, then roll the end of the wire around

FIG. 6.

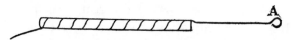

Appearance and size of wire and loop, *A*.

another match until you have bent the end of the wire into a small loop (Fig. 6). Prepare a little powder of common borax, and then, heating the wire loop in the general flame, plunge it quickly into the powdered borax. It will immediately pick up a quantity of the powder, and then, by turning the flame upon the borax, you will have a clear and perfectly transparent bead filling the little loop on the end of the wire. You are now ready for the

3

experiment of illustrating the special properties of the two flames, which we shall now describe.

Obtain some black oxide of manganese, from any druggist, and dropping a little upon a clean sheet of letter paper, heat your borax bead red-hot in the flame and quickly touch with the hot bead a particle of the black oxide—it will stick to the bead—then turn the outer or O flame upon the bead and blow till the particle of oxide of manganese has entirely dissolved—it will impart to the bead a beautiful amethystine-purple. Now turn the inner flame, that is, the R flame, upon the bead, and in a few seconds (according to skill in keeping the R flame steadily on the bead) the color will disappear, but it will return when the O flame is used again.

These efforts will give practice, ending in sufficient skill to enable the learner to use the blow-pipe as directed in the future parts of this work.

The various reactions of different substances are given in the body of this book as they are called for when the substances are described.

A glass tube of a little less than three-eights of an inch in diameter may be made into a blow-pipe as follows: Take a piece of such a tube, ten or twelve inches long, soften the tube by red heat in an alcohol flame, and draw it out to a small diameter—cool and scratch or file it at the smallest diameter —break it off, introduce the tube into the flame again and bend the glass to a right angle, about two inches off from the point—cool gradually—and heat the mouth end, opening it a little by introduc-

ing a small dry pine stick, cool it, and you have a
very efficient blow-pipe when another of metal cannot be had.

Note: If your platinum loop will not hold the
borax bead, then it is too large. Make a smaller
loop. If it is dimmed or blackened by smoke, heat
it red-hot—it will clear up.

The three principal means of chemically testing
minerals before the blow-pipe are (1) with borax ;
(2) on charcoal, usually with the addition of carbonate of soda ; (3) by holding in the oxidizing
point.

In connection with this the following experiments
given by Alexander M. Thomson, D. Sc., are of interest :

Experiment. No. 1.—Many metals impart a color
to fused borax, by which their presence can be
recognized. To try this experiment, a bead of
fused borax must first be obtained on the platinum
wire. The end of the wire is bent into a loop or
ring about the twelfth part of an inch in diameter.
The wire is then heated in the blow pipe flame, and
dipped whilst hot into the borax ; the portion of
borax that adheres is then fused on to the wire in
the blow-pipe flame, and the hot wire is again
dipped ; this is repeated until the loop contains a
glass-like bead of borax. If the bead has become
cloudy, the soot causing this may be burnt off in
the oxidizing point of the flame. Having thus obtained a clear, colorless, transparent bead, the next
step is too add to it a minute portion of the mineral

which is to be tested. By touching a little of the finely-pulverized mineral with the borax bead, while softened by heat, enough will adhere to the bead for a first trial. The bead is then kept at a white heat in the oxidizing point of the flame for a few seconds, and on removal its color is noted, both whilst hot and when cold. If no color is imparted, a fresh trial may be made with a larger quantity of the powder; but if the bead is opaque owing to the depth of color, as is often the case, a fresh experiment must be made, using a still smaller quantity of the powder. The color can only fairly be judged in a perfectly transparent bead. If no color can be obtained in the oxidizing point, further experiment with the borax bead is needless; but if a color is obtained, it is then advisable to try the effect of the reducing flame upon the same bead. The following observations and inferences may result from this test :

COLOR OF BEAD IN

Oxidizing.	Reducing.	Presence of.
Green (hot); blue (cold) .	Red	Copper.
Blue (hot and cold)	Blue	Cobalt.
Amethyst	Colorless	Manganese.
Green	Green	Chromium.
Red or yellow (hot) . . . ⎫ Yellow or colorless (cold) ⎭	Bottle-green	Iron.
Violet (hot); Red-brown (cold)	Gray and turbid, difficult to obtain .	Nickel.

This mode of testing may often be used to prove the presence of the above-mentioned metals.

It requires some practice before reliable results can be obtained in reducing. The reduced bead, if brought out of the flame at a white heat, into the air, may at once oxidize; but this may be prevented by placing it inside the dark inner cone of an ordinary candle flame, and allowing it to cool partially there.

Experiment No. 2.—the mode of testing with carbonate of soda on charcoal, is performed as follows : A sound piece of charcoal half an inch square is chosen, and a neat cavity is scooped out on its surface, into which is placed a mixture containing the pulverized mineral to be tested, with three or four parts of carbonate of soda, the whole not exceeding the bulk of a pea. After lightly pressing the mixture into the cavity, the blow-pipe flame may be cautiously applied to it; and afterwards when the mixture no longer shows a tendency to fly off, the charcoal may be advanced nearer to the blow-pipe, and finally be kept at as high a temperature as possible, in the reducing part of the flame.

In testing for tin ore, a piece of cyanide of potassium, about the size of a pea, may be placed upon the mixture after the first application of heat, and the further application of heat may then be continued.

This treatment is designed to extract metals from minerals ; it favors in the highest degree the removal of oxygen. But like the borax test, it is limited in its application, as it can only be used to detect certain metals. The failure of the test in any case must not be looked upon as a conclusive proof of the absence of the particular metal sought; for instance, copper can be easily extracted from carbonate of copper by this test, but not from copper pyrites. Still the test is a most valuable and indispensable one to the mineralogist. The test is complete when the metal is obtained as a globule, in the cavity of the charcoal. In many cases the globule will be found surrounded by the oxide of the metal, forming an incrustation on the charcoal; and the color of such incrustation should be carefully noted, both at the moment of removal from the flame, and after cooling. By pressing the globule between smooth and hard surfaces, it can be determined whether the metal is flattened out (or malleable), or crushed to pieces (brittle).

The following observations and inferences may result from this test :

Globule.	Incrustation.	Presence of.
Yellow, malleable	None.	Gold.
White, malleable	None.	Silver.
Red, malleable	None.	Copper.
White, malleable	White	Tin.
White, malleable	Red (hot); Yellow (cold)	Lead.
White, brittle	Red (hot); Yellow (cold)	Bismuth.
None	Yellow (hot); White (cold).	Zinc.
White, brittle, giving off fumes when removed from the flame	White	Antimony.

Experiment No. 3.—In addition to these substances there are others which occur abundantly in minerals, and which may be recognized by the blow-pipe with the greatest ease; for instance, sulphur and arsenic. These may be discovered by heating a fragment of the mineral, supported on a piece of charcoal or held in a forceps in the oxidizing point of the flame, and comparing the odor which is given off. A smell of burning sulphur indicates that the mineral contains that substance, and white fumes having a garlic odor indicate the presence of arsenic.

Mercury, antimony, and other substances may escape as fumes when heated in this manner.

Nitrate of cobalt dissolved in water, and used in exceedingly small quantity, helps to discriminate between certain white minerals, such as kaolin, meerschaum, magnesite, dolomite, etc. The mineral is reduced to powder and moistened with a drop of a very light solution, and then heated before the oxidizing flame of the blow-pipe. Kaolin and other minerals containing alumina assume a rich blue color, while meerschaum and other minerals containing magnesia become flesh-colored. Oxide of zinc, under the same circumstances, becomes green, and this can be tried with the white coating obtained on charcoal by reducing an ore of zinc with carbonate of soda.

Tests in glass tubes can be better made over a spirit lamp, so as to avoid the deposit of soot on the glass, but they can also be made with the blow-pipe flame, provided it is used carefully, avoiding too sudden a heat, which would break or fuse the glass. The presence of water in minerals will be detected in this way, and the water collects in small drops in the cold part of the tube. Some minerals containing sulphur, arsenic, antimony, tellurium and selenium often give a characteristic deposit.

Minerals containing mercury can also be tested in this way, as by adding a little carbonate of soda, sometimes with cyanide of potassium, a sublimate of metallic mercury will be formed in the cold part of the tube. A little charcoal should be added to arsenical minerals.

Organic combustible minerals generally leave a deposit of carbonaceous matter at the bottom of the tube, and the volatile hydrocarbons condense in the cooler part. The tube should, therefore, always be long enough to allow for this condensation. Minerals which yield a characteristic smell will be best tested in this way.

CHAPTER III.

THE *forms* which many minerals assume always indicate their *composition.* It is, therefore, sometimes a great help to the prospector to become acquainted with the subject of crystallography so far as to enable him to determine the system or order to which a crystal belongs.

We shall treat of the subject only so far as may be of practical application to the purposes of the prospector in the search for the useful minerals.

It is necessary to understand that nearly all mineral substances, when they appear in the crystalline condition, assume a characteristic form and do not trespass upon that of other minerals. Although, to the unaided eye and unskilled vision, this assertion may appear to be a mistake in some few cases, it appears so only because the differences are exceedingly small.

All crystalline forms have been reduced to six classes or systems, which are named as follows : I. *Isometric;* II. *Tetragonal;* III. *Hexagonal;* IV. *Orthorhombic;* V. *Monoclinic;* VI. *Triclinic.*

I. ISOMETRIC system. The principal forms of this system are the cube, octahedron, dodecahedron, the two trisoctahedrons, the tetrahexahedron, and the hexoctahedron.

(41)

The *cube* has six equal and square sides, as in Fig. 7. In this form lines drawn from the centre of each face to the face opposite, cross each other at *right angles*, and are of the *same length*.

This system is called *isometric*, that is, iso *equal*, and metric *measure*, because these axes or lines are of equal length and at right angles to each other. It must, however, be remembered that the cube is modified in some minerals, but wherever these modifications take place the original form of the cube may always be traced. Some of the changes may be very intricate, and these especially unusual or intricate forms we shall not notice. The usual forms only are of importance, and can be treated of in so small a work as this.

The learner should take a potato and cut as perfect a cube as possible, and make himself acquainted with the common variations which may belong to the cube, as we shall show, without changing the length of the axis, and always, cutting so that the axis will always be the same or of equal lengths.

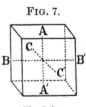

FIG. 7.

The Cube.

Fig. 7 is the cube with the three axes $A A'$, $B B'$, $C C'$. If, with your knife, you slice off one edge angle from A to C' and from A to C, and in like manner from A to B' and from A to B, you will have a four-sided pyramid, the apex of which will be at A and the four-sided base at $C B'$, $C' B$, or around one-half the cube. Now, treat the opposite side in the same way, and you will then have the following figure, which is the octahedron (Fig. 8).

The dodecahedron (12 sides), Fig. 9, may be formed by taking off the solid angles *A*, *B*, *B'*, *A'*. In all three cases and many others, the three axes remain the same in length and in their angular direction where the forms have not been distorted.

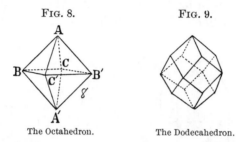

FIG. 8.   FIG. 9.

The Octahedron.   The Dodecahedron.

II. TETRAGONAL system. The chief forms of this system are the two square prisms and pyramids, and the eight-sided prism and double eight-sided pyramid.

The tetragonal system has also three axes as in the isometric, and they are at right angles to each other, but the vertical axis is longer than the others, as in Fig. 10.

The term *tetragonal* means " four-cornered or angled," and is not precise, for a cube is tetragonal, but it is used to express this form because it is one word; otherwise "square prismatic" would be a more correct description, since Fig. 10 is that of a prism; for in mineralogy any crystal having parallelograms for sides is called a prism. Cut this prism as in the case of the cube, and you will have the form seen in Fig. 11.

Variations upon this form may show a prism with four-sided termination at either or both ends, as in Fig. 12. This is the form of the transparent gem called the zircon, anciently called the jacinth. The zircon has been mistaken for the diamond, which it resembles in brilliancy, and somewhat in hardness. But the diamond is isometric and never tetragonal,

FIG. 10.   FIG. 11.   FIG. 12.

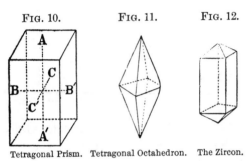

Tetragonal Prism.   Tetragonal Octahedron.   The Zircon.

and hence it may be distinguished readily from the zircon.

III. HEXAGONAL system. The chief forms of this system are the two six-sided prisms, the two double six-sided pyramids, and the twelve-sided prism and double twelve-sided pyramid. It differs from the tetragonal system in that it has three equal lateral axes instead of two; the vertical being at right angles, as in Fig. 13, with each of the three lateral.

But it must be remembered that owing to various causes in nature the hexagonal crystal always calls for hexagonal terminations; thus Figs. 14 and 15.

Owing to various causes in nature, the hexagonal crystal may be found under various modifications of the hexagonal form, but it can always be reduced to this system. The symmetry of the crystals may be by sixes, or very rarely, by cutting each angle it may be in twelves, or the sides may be unequal in area or length, as in Fig. 14. The author once found a quartz crystal in Switzerland which was, for nearly its entire length, three-sided, but showed its hexagonal nature only at the extremity, where, hav-

FIG. 13.   FIG. 14.   FIG. 15.

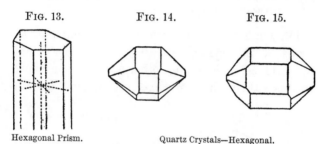

Hexagonal Prism.   Quartz Crystals—Hexagonal.

ing been free from its confinement in process of formation, it had assumed its normal crystallization. As has been said in another place, calcite crystals sometimes assume a hexagonal prism precisely as does quartz, but the latter shows always six-sided terminations, whereas lime or calcite crystals show three-sided terminations, as in Figs. 16 and 17. There are two sections or forms of this system, the *hexagonal* and the *rhombohedral;* both belonging to the hexagonal system, and distinguished as we have shown.

These calcite crystals belong to the rhombohedral

section of the hexagonal system, showing rhombo-hedral forms at the end, as in Fig. 11.

FIG. 16.

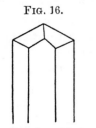

Calcite hexagonal crystal—three-sided termination. Side view.

FIG. 17.

The same—end view.

IV. ORTHORHOMBIC system. The characteristic forms of this system are the rhombic prism and pyramid. There are also other forms called domes. In this system the three axes are unequal and in-

FIG. 18.   FIG. 19.

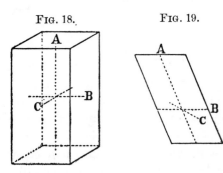

tersect at right angles, as in Fig. 18, wherein the axes, A, B, C, are unequal in length, but at right angles at the intersection. The terminations are flat although frequently beveled on the surrounding edges.

V. MONOCLINIC system. The monoclinic forms are too difficult to be fully described here, but it is not hard to learn what is most essential about them. In this system two of the axial intersections are at right angles; but one is oblique, and the side of the crystal is inclined as in Fig. 19.

Crystals of feldspar in general which contain potash (called orthoclase or potash feldspar), are monoclinic, but the soda feldspar crystals belong to the next or sixth system, as do also the lime feldspars.

VI. TRICLINIC or "*thrice inclined*" system. In this system the planes are referred to three unequal axes all oblique to each other. The only important feature in this system is that there is no right angle in any of its crystals; but it is of little use for our purposes, since with the exception of the lime feldspar and soda-lime feldspars (anorthite or lime feldspar, labradorite or soda-lime feldspars, andesite and oligoclase, both soda-lime feldspars, and albite, a soda feldspar), all the rest are of little importance, except microcline, a potash feldspar.

As ILLUSTRATIONS OF THESE SYSTEMS the following may be stated :

Of the isometric system, or FIRST system, are gold, silver, platinum, amalgam, copper, the diamond, garnet, magnetite, pyrite, galena, alum, kalinite, all of which assume the cubic octahedral, or some allied form.

Of the tetragonal, or SECOND system, are the zircon, chalcopyrite, cassiterite (tin ore), titanic oxide, and others.

Of the hexagonal, or THIRD system, are beryl, aquamarine, the emerald, chrysoberyl, apatite (lime-phosphate), quartz.

Of the orthorhombic, or FOURTH system, are, barite or sulphate of barytes, celestite or sulphate of strontia, and carbonate of strontia, also cerussite or lead carbonate.

Of the monoclinic, or FIFTH system, are borax, gypsum, glauber salt (*mirabilite* is its mineralogical name), copperas (or *melanterite*).

Of the SIXTH system we have already given sufficient illustrations.

Of the GEMS not mentioned in the above, the TUR-QUOIS owes its blue to copper, and is never crystallized, being in reniform or stalactitic conditions. It is a phosphate of alumina with water in composition. This mineral or gem should be carefully distinguished from LAZULITE, which, though blue, crystallizes in the *monoclinic*, or fifth system ; it is a softer mineral and contains considerable magnesia, lime, and iron, of which (except a very small amount of iron), the true turquois contains none. The latter is the gem, and may be beautifully polished, and keeps its color, which is due to copper. Lazulite is found in beautiful crystals at Crowder's Mount, in Lincoln Co., N. C.; also fifty miles north of Augusta, at Graves's Mount, in Lincoln Co., Georgia.

Both these should also be distinguished from LAPIS LAZULI, which also crystallizes, but in the isometric or first system, though commonly massive and compact. This is valuable in the arts, and when powdered forms the *ultramarine*, a rich and durable paint. It is a silicate of alumina, but contains some lime and iron. It is used also for costly vases. But the artificially prepared ultramarine is largely used in the arts. The native mineral is found in syenite and in metamorphic crystalline limestone, associated with pyrite and mica.

The TOPAZ crystallizes in the orthorhombic section of the hexagonal or fourth system. The finest are generally in prismatic form, showing a flat plane at the extreme end, even when the end of the crystal has several inclined faces. It is a silicate of alumina with fluorine. The fluorine may be detected before the •blow-pipe in the open tube by powdering a little of the topaz and mixing it with a little microcosmic salt (a salt of phosphorus). The heat of the blow-pipe will let free the fluorine, and its strong pungent smell, and its corrosion of the tube, will prove its presence. With the cobalt (nitrate) solution on charcoal, it gives a fine blue color in proof of alumina. This is the best test of the topaz, as the color of the mineral is not always the same, nor is it always perfectly transparent. It is found at Crowder's Mount, already spoken of, and also in Thomas's Mountains, in Utah, near lat. 39° 40′ and long. $113\frac{1}{2}$° W. west of south of Salt Lake (Dana). In Trumbull, Conn., the crystals are abundant, but not very transparent.

METEORIC IRON has been reported from North Carolina as found native in a partial crystal of the

4

isometric form, and several meteoric masses from Arizona have been reported at the Geological Section at Washington, D. C., September, 1891, as containing black diamonds, small but interesting.

Meteorites are less pure than native iron, the iron in them being almost invariably associated with nickel, and they also contain traces of cobalt, copper and other metals. In the many specimens examined, the iron ranges from 67 to 94 per cent., and the nickel from 6 to 24. Their masses generally range from a few pounds in weight to a ton or more. If cut, and the surface is polished, and then acted upon by nitric acid, a kind of etching action goes on, the acid acting on spaces between bands of untouched metal which cross the mass in two or three directions, and in these the nickel is more abundant than in other parts, for it is not equally diffused in the alloy.

RUBY AND SAPPHIRE. These crystallize in the rhombohedral form.

The garnet is sometimes mistaken for the East Indian ruby, which is the most precious variety, but the garnet is *isometric*, and even when cut and mounted may be distinguished from the oriental ruby by the superior hardness of the ruby, the latter being next to the diamond, while the garnet is only as hard as quartz, or not quite so hard. So that a garnet of the most precious kind if worn will, under the strong lens, show the lines of wear, especially on the edges, which are absent in the true oriental ruby. Oriental garnets are frequently confounded with rubies by jewelers in Paris as well as in America. For instance, some years ago, two oriental garnets worth about $20 each were found to be set in a diamond ring as oriental rubies, for which the sum of $2,000 was paid. The firm in Paris acknowledged the mistake, and refunded the $2,000. The oriental ruby is essentially pure alumina, while the oriental or precious garnet is a silicate of alumina with lime and a little iron.

All these gems are found in the crystalline rocks, as granites, gneiss, dolomite, and some (topaz, ruby) associated with tourmaline, tin ores, mica, etc., and the crystalline lime-stones. The true turquois is found in Persia in the clay slates in veins running in every direction. Very good specimens have been found in Arizona and New Mexico; also in Colorado in the Holy Cross Mining district, thirty miles from Leadville.

CHAPTER IV.

THERE are a few simple measurements which are sometimes desirable, and which can be made without the labor of carrying instruments and chains. The actual work of surveying, to be of any value to the prospector, must be so accurately performed that the work should be entered upon as a specialty, and he must use a theodolite or transit and make use of logarithms. Any small work on surveying or trigonometry will give sufficient information.*

Some few measurements, however, and simple surveys with easy methods, are given here to meet cases where only a general approximation is required.

TO MEASURE HEIGHTS WHICH ARE INACCESSIBLE.

Any height of tower, stand-pipe, tree, etc., may be measured approximately by knowing your own height and taking advantage of sunlight, thus:

Let $A B$, Fig. 20, be the height of the object to

---

* For this purpose we would recommend the following book : The Practical Surveyor's Guide. By Andrew Duncan. A new, revised and greatly enlarged edition. Illustrated by 72 engravings. Philadelphia, Henry Carey Baird & Co., 1899. Price, $1.50.

be measured. The dotted line is the shadow cast. Walk off into the sunlight and note on the ground the point at which your own shadow terminates; measure from the heel to that point. A calculation in single "rule of three" will give $A B$ thus:

$$C' B' : B' A' :: B C : A B.$$

Heights of hills or land may be nearly enough measured by the aneroid barometer, the instructions in the use of which go with the instrument, or may be obtained with it, and approximately accurate aneroids may be had small enough to go into the side pocket, or still more accurate ones may be easily carried in a case held by a small strap around the shoulders. For hills under 2000 feet, the fol-

FIG. 20.

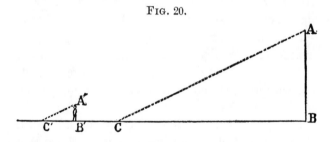

lowing rule will give a very close approximation, and is easily remembered, because 55°, the assumed temperature, agrees with 55°, the significant figures in the 55,000 factor, while the fractional correction contains *two fours*.

Observe the altitudes and also the temperatures

on the Fahrenheit thermometer, at top and bottom respectively of the hill, and take the mean between them.   Let B represent the mean altitude and b the

mean temperature.   Then $5500 \times \dfrac{B - b}{B + b} =$ height

of the hill in feet for the temperature of 55°.   Add $\frac{1}{440}$ of this result for every degree the mean temperature exceeds 55°; or subtract as much for every degree below 55°.

TO MEASURE AREAS.

Theoretically, it is very easy to "step off lines," but practically it is very difficult thus to arrive at accuracy on uneven land.   But where one is acquainted with the exact average measurement of his step on level land, he may reach some approximate accuracy on uneven land by remembering that in ascending, even slightly, his average decreases, and *vice versa* in descending.   A good strong tape measure, kept on a level in ascending and descending hills, is more convenient and more easily handled than a chain.

1. On square areas the length of the side multiplied into that of the adjacent side gives the area.

2. In the parallelogram, where all angles are right angles, the same is true.

3. In any other shapes the following rules are to be observed :

First: Measure the area of a right-angled triangle thus :

Let *B*, Fig. 21, be the right-angle; the area of *A B C* is equal to the length,
B C, multiplied into half the perpendicular distance, *A B*.

FIG. 21.

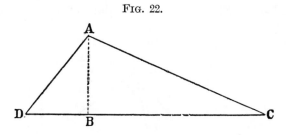

Example: *B C* = 100 ft.; therefore, if *A B* = 90 ft., 100 × 45 = 4500 sq. ft. = area of *A B C.*

The same rule applies when the triangle is not a right-angled triangle; thus, the angle at *A*, Fig. 22, being obtuse.

D C = 150 ft., *A B* = 90 ft.; multiply 150 ft. by

FIG. 22.

one-half *A B* = 45 ft., and we have 6750 sq. ft., for *A C D* is composed of two right-angled triangles, *A C B* and *A B D*, as in the previous example.

Or, when the triangle has an acute angle at *A*, Fig. 23, thus: Treat precisely as in Fig. 22, only letting the perpendicular fall from *D* upon *A C*, that is, invert the triangle.

The cases wherein the sides are more than three

are treated by resolving all such areas into right-angled triangles, thus:

In Fig. 24, the area, *A C D B* may be resolved into two triangles, *A C B* and *C D B*, of which *A*

FIG. 23.

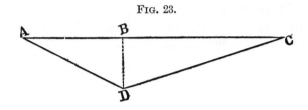

B is the base of the one and *C B* that of the other. In Fig. 25, the area, *A C D B E K*, may be resolved into the four triangles, *A C D, A D B, A B E* and *A E K*. The perpendiculars of Fig. 24 are

FIG. 24.

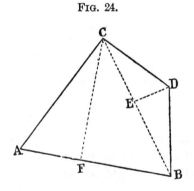

E D and *C F*. Those of Fig. 25 are *C H, I B, F E*, and *K G*, and the length of bases may be multiplied into half that of the perpendiculars, as in the case already given, and the feet be reduced to acres, rods, etc., or miles.

For the number of square feet in an acre, etc., see Appendix No. 3, and treat it thus: Suppose the area of Fig. 25 be 80,000 sq. ft., then according to

FIG. 25.

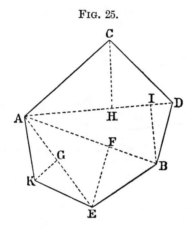

Table No. 3, it will be 1 acre, 3 rods, 13 poles, 25 yards, 7 feet, or 1.836 + acre.

TO MEASURE AN INACCESSIBLE LINE.

Suppose we desire to measure the distance across a river, as in Fig. 26.

We want to find the distance *A B*. Measure a distance of about 100 ft., *B D*, at right angles to *A B*, and raise a pole at *C*, about half-way from *B* to *D*. Proceed in measuring at right angles to *B D*, in the direction *D E*, letting *E* be that point at which the line *C E*, if extended, would strike *A*.

Now you have two right-angled triangles of the
same angles, for, as every triangle has two right
angles according to geometry, and each of these triangles has one right angle, and the opposite angles
at C are equal according to geometry, the remaining

FIG. 26.

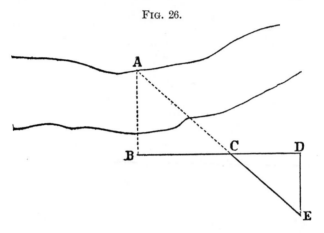

angles at A and E are equal, and the triangles are
proportional, and the proportion is—

$$C\,D : D\,E : : C\,B : A\,B.$$

Then, if $C\,D = 40$ ft., $D\,E = 45$ ft., and $C\,B = 60$,
we know that $45 \times 60 = 2700$ divided by $(C\,D)$ 40
ft. $= 67\frac{1}{2}$ ft.; this is for $A\,B$, or the distance across
the river.

The only difficulty is in measuring your angles
as true right angles, and this may be done by
measuring the perpendicular, thus—

Extend the line $A\,B$, Fig. 26, to F, Fig. 27, and

likewise the line *D E*, Fig. 26, to *C*, as in Fig. 27.
Now measure equal distances on the line *B D*, for
the lines or offsets, *B C* and *B H;* also from *D C*,
the offsets *D I* and *D K;* drive sticks in at *G, H,
I,* and *K.* See that the distances represented by the
dotted lines are equal, and if so, the lines *A B F*
and *D C* are perpendicular to the line *G K*, and

FIG. 27.

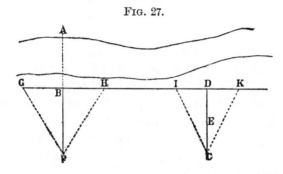

your work will be well done and very nearly accurate.

It is, however, well for the prospector to use a
prism˙ compass which will read to one-quarter degree. Such a compass may be had at very low rate,
not more than three inches diameter, of light
weight and of sufficient accuracy. The author has
used one for many years, and traveled with it many
thousands of miles in Asia and Africa, and can
testify to the fact that by customary use it may be
handled to a great degree of accuracy for horizontal
angles. The needle is attached to the under side of
a cord with steel engraved degeees and fractions,
and read by a magnifying prism.

In almost every conceivable surveying project, especially in running adits and sinking shafts to strike adits and galleries, only the best instruments should be used. Everything depends upon the most accurate measurements, and this department of engineering is not one that can be treated approximately, because any error in measurement may result in very provoking and expensive mistakes.

We have presented all that is required on surface measurements, except where it becomes necessary to make such accurate proceedings as may only be executed by use of the finest instruments, and that with considerable practice. Otherwise accurate mathematical tables are of little importance, as their use is based upon the presence of most accurate data, and without this the best methods and diagrams are in vain.

This subject of mining engineering does not come within the range of our work, and for all mere exploring as a prospector such ground-work or digging for examination as is necessary will readily suggest itself to any intelligent workman.

CHAPTER V.

I. *Wet Method.*

PRELIMINARY EXAMINATIONS may be made at first with the pocket lens and a piece of steel or a heavy-bladed pocket-knife. The *first*, to see if any native metals or any sulphides, etc., are present; the *second*, to try the softness or silicious nature of the mineral; if much quartz (silex) is present it will strike fire.

Pulverize a small part and use the blow-pipe to detect SULPHUR, ARSENIC, SELENIUM, by the smell on charcoal or in the glass tube. Arsenic fumes have a garlic odor, silenium that of horse-radish.

Use a test tube with a little nitric acid and heat over a spirit flame. Add a few drops of water and one drop of sulphocyanide of potash—an intense deep red appears, deeper according to amount of IRON and solvency of the mineral in nitric acid.

Try another portion in the same way, but drop one drop of hydrochloric acid. A dense curdy white precipitate indicates SILVER.

Native gold or silver is determined by color and softness, as we have elsewhere stated (*see Index*). Treat another portion in the same way with nitric

(61)

acid, drop in several drops of strong ammonia water. The blue color indicates COPPER.

Antimony and *tin* are detected by the blow-pipe. Place the former upon charcoal with carbonate of soda, and brilliant metallic globules are obtained; the metal fumes and volatilizes, and covers the charcoal with white incrustations, and needle-shaped crystals appear. *Tin* appears when the ore is mixed with carbonate of soda and cyanide of potassium on charcoal, and the inner flame turned on—ductile grains of metallic tin and no incrustations appear.

Manganese gives amethystine beads of borax in the outer flame, *O F*, disappears with the inner, *I F*, reappears with the *O F*.

Alumina, magnesia, lime, give their characteristic colors, or in the last case, incandescent light before the blow-pipe on charcoal. *Alumina* heated on charcoal, and then touched by a half drop of proto-nitrate of cobalt, then heated strongly in the *O* flame, gives a blue color. *Magnesia* so treated gives a faint red or pink, seen just as it cools.

Zinc heated on charcoal with carbonate of soda in the reducing flame becomes metallic, and when oxidized in the *O* flame gives a white oxide which is yellow when hot, white when cooled, and with proto-nitrate of cobalt when heated in the *O* flame, a beautiful characteristic green color.

Cobalt and *nickel* give the colors we have noticed in another place under their respective names (*see Index.*)

Uranium heated with microcosmic salt (phosphate of soda and ammonia), on platinum wire in the *O* flame dissolves, producing a clear yellow glass, which, on cooling, becomes yellowish-green. But the analyst should remember that copper also produces a green bead, but *only* in the outer or oxidizing flame, and *chromium* the same, but in both outer and inner flames.

The *copper* green becomes blue on cooling, the *chromium* green remains green on cooling. This will always prove the metal.

Titanium in the presence of peroxide of iron, as in some titanic ores of iron and sand, gives, with microcosmic salt in a strong reducing blow-pipe flame, a yellow glass, which on cooling becomes red.

Mercury may be detected in almost any of its ores by the process described (*see Index*), by heating in a glass tube and noting, under the lens, the sublimation of mercury in very minute shining particles.

Minerals which are *carbonates* may be detected by their effervescence when touched by a drop of hydrochloric acid, as in limestone and spathic iron ore. But the analyst must remember that some cyanides effervesce where neither lime nor carbonic acid is present, and chloride of lime where there is no carbonic acid. With these latter other tests must be used, but the sense of smell will show that carbonic acid does not exist, the latter having no odor.

Some sandstones have a small amount of lime carbonate and must be tried under the lens, as the

bubbles are minute.   But, while in these examinations great help is received, and many determina-
tions made, especially in simple minerals and ores,
there are compound ores so mixed in elements that
the above tests fail to give satisfaction, because the
colors are mixed and the action confused.   Some of
the elements must be moved out of the association
and a separation made.   This analysis is called
qualitative, and we shall take a case of very full
analysis of a compound ore.

QUALITATIVE ANALYSIS OF ORES where many elements are present :

There are many times when it becomes not only
a matter of curiosity but of importance for the prospector to know the entire composition of the ore he
has before him.

With a little practice the "wet method," as it is
called, may be used by the prospector with all the
accuracy required under the circumstances.

The "dry method" of analysis is that in which
no liquids are used, but only fluxes and heat.
Although for one or two elements it is simpler than
the wet method, it may so happen that sufficient
heat cannot be had.   We shall, therefore, give some
directions whereby the wet method may prove of
greater service.

1. Pulverize the ore as finely as possible and
sieve it, passing the entire· quantity taken as an
assay.   Should any part be left remaining in the
sieve it may be a very important part.   Pass the
whole through.

2. Take a test tube and drop a little of the sifted ore into it, pour a little nitric acid upon it, add about one-eighth part water, warm it gently over a spirit flame to see if it will dissolve; if not, then add four times as much in bulk of muriatic acid (hydrochloric acid). If this will not dissolve then proceed as follows:

3. Put the assay, after fine pulverization, into a platinum crucible. Place it in a suitably arranged platinum wire triangle so that it will hang over an alcoholic blast lamp. When all is ready add a mixture of equal parts of sodium carbonate and of potassium carbonate, amounting in all to about four times the bulk of the assay, stir gently with a glass rod or a stiff platinum wire, and then light the lamp. Watch the assay, and when it begins to swell up withdraw the lamp, but return it when the swelling subsides, so that the alkalies do not throw your assay out of the crucible, which should be only one-half full at the beginning. With care the contents will soon subside, and under increased heat become a quiet liquid mass. Now, extinguish the flame, cool the crucible, remove crucible contents to a beaker glass or place the crucible with its contents within the beaker, and pour a little water upon it, add some nitric acid, or a little hydrochloric acid, *but not the two acids together*, unless you have only the assay and not the platinum crucible in the beaker—nitro-muriatic acid dissolves platinum. Warm and stir till the assay is entirely dissolved, except perhaps some white grains of silex.

5

4. If the preceding work has been properly performed, the assay is now dissolved and you are
ready for work. Filter the contents of the beaker
to separate any undissolved remainder, if any such
is seen in the glass, and wash the filter-paper by
passing an ounce or two of water through it, and
now make preparations for the next step. It is not
necessary, where extreme accuracy is not required,
to wash the filter-paper perfectly free from the acids.
But if it be necessary, then furnish yourself with a
small strip of platinum ribbon ; clean its surface to
a polish. If a drop of the filtrate evaporated from
this surface shows not the least trace of sediment or
outline even under a lens, the filter-paper is sufficiently washed. When the filter-paper is to be
burned and weighed, it must be perfectly freed from
the acids by continuous washing.

5. Pour ten or fifteen drops of the filtrate into a
test tube. Drop in three or four drops of hydrochloric acid. If a precipitate forms it may be of
SILVER; if so, it will grow dark violet on exposure
to daylight, or more rapidly and darker in sunlight.
Or to test more quickly, add strong ammonia, 30 to
40 drops ; it dissolves after a short time; or if it does
not dissolve then it is LEAD ; filter and test on
charcoal with the blow-pipe; if it gives, with inner
flame, a bead and yellow incrustation around, it is
LEAD. Or, if none of the above results are seen, and
yet there is a precipitate, then it is *mercury.* To
prove this, add a solution of carbonate of potash and
digest ; it turns black ; filter and place it in a glass

tube, heat gently with a blow-pipe; it volatilizes and condenses on the sides, examine with strong lens, it is *mercury*.

6. But suppose hydrochloric acid produces no precipitate though in excess and heated? Then there is neither LEAD, SILVER nor MERCURY in the assay, and it is not necessary to treat the ore for either, but proceed to the next step. It will be seen why we directed nitric acid to be poured on the assay, as in No. 2. Hydrochloric acid would have prevented these tests as given, but you are now prepared for the next metals, with three less to look for, or with a certainty as to the presence of one or more of the three.

7. The whole assay, or its solution, may now be used. If any precipitate occurred in the test tube, treat the whole assay solution with hydrochloric acid, heat to boiling, and separate the precipitated metal or metals in the whole, as in the test tube, by filtration. Wash, set the paper (filter) aside under cover of paper to dry, and pass hydrogen sulphide slowly through the filtrate until the filtrate smells plainly of the gas.

8. As this gas is frequently used, make a simple and cheap apparatus so that you may have a supply at any time, thus: Cut off the bottom of a long bottle * of small diameter, D, say about two inches, and fit it into a fruit jar, E, as in Fig. 28.

* Cut a nick. with a large file. in the spot where you wish to start a crack near the bottom, then heat a rod, or poker, or spike-nail, nearly red-hot, place it on the nick, a crack

The top *A* should be fitted loosely so that it may be removed and let air pass through. The cork at *B* must be air-tight. Fit a small tube into the cork after bending it in a spirit-lamp flame—a quarter-inch tube with an eighth-inch aperture is sufficiently large and is easily bent. Take an inch rod of iron, let the blacksmith heat it white-hot, and press it into a small roll of brimstone, this will give you iron sulphide—you need it in pieces as large

FIG. 28.

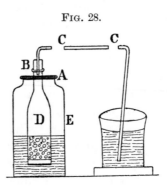

as bullets: it melts readily against the brimstone. Place some cotton in the neck of the bottle, and having fitted a plug of wood with holes in it for the bottom of the bottle, invert the bottle and fill it half full of iron sulphide lumps, fasten the wooden plug in the bottom, not very tightly, but tightly in three or four places, so that water can pass easily,

starts; draw your hot iron and the crack will follow; when nearly cracked around pull the bottom off. A glass chimney may be used, but it is rather too small to contain sufficient iron sulphide.

and yet the plug be well fixed in. Put the bottle in its place, resting in the jar at *A*, and somewhat loosely fastened. But this must be after you have half filled the jar with a mixture of equal parts of common hydrochloric acid and rain-water (or, next best, well-water). Hydrogen sulphide will form immediately, and if you have made all connections perfectly, as in the figure, the gas will pass from this apparatus into the solution of ore in the beaker and precipitation will soon take place. The advantage of this apparatus is that if you tie two little blocks of wood against the sides of the India-rubber tubes, *C C*, so as to press the sides together and stop the gas from flowing, the gas forming pushes the water out of the interior glass *D*, and the gas stops forming, but is ready at any moment to begin as soon as the string around the little blocks is removed.

9. After introducing the hydrogen sulphide until the filtrate smells of the gas, filter and wash the precipitate, mark the paper containing it with the letter *A*, and put this precipitate aside for the present. This is the *precipitate* from the hydrogen sulphide.

10. THE FILTRATE. If the strip of platinum shows that it contains some material after evaporation of a few drops, proceed by adding a solution of ammonium chloride (sal ammoniac), and then aqua ammonia to the filtrate, using about one-fifteenth or one-twentieth of the. bulk. Then add ammonium sulphide so long as any precipitate is apparent. Let it stand awhile. This precipitate may

contain alumina, chromium oxide, zinc, nickel, manganese, cobalt and iron as sulphides. It may likewise contain phosphates, borates, oxalates, and hydrofluorates of the alkaline earths (barium, strontium and lime). The latter we may not care for.

11. Filter and wash this precipitate. Add a little water to the hydrochloric acid, now to be used in treating this precipitate. Add this diluted hydrochloric acid in sufficient quantity to dissolve the precipitate, and put it aside to digest. If any part refuses to dissolve, it is because there may be present cobalt, or nickel, or both ; add nitric acid and boil, for these metals dissolve in hot nitro-hydrochloric acid. Filter. Next add to the whole solution ammonium chloride, and excess of aqua ammonia. The consequent precipitate may contain alumina, chromium oxide, sesquioxide of iron, and the alkaline earths, as phosphates, etc. Dissolve the precipitate by digesting in caustic potash solution till all is dissolved that will dissolve. Filter. The solution may contain alumina and chromium oxide ; boil for some time, and if a precipitate is formed, it is CHROMIUM OXIDE ; confirm by the blow-pipe ; it gives a green bead with borax, heightened by fusion with metallic tin or charcoal, which is the blow-pipe test for *chromium.*

12. Now super-saturate the solution with hydrochloric acid and boil with excess of ammonia ;* if a precipitate is formed it is *alumina.* Confirm with

---

* By "excess" is meant so much that after stirring with a glass strip or rod, the liquid smells strongly of ammonia.

blow-pipe, as we have shown. What was dissolved by digestion with potassium hydroxide (caustic potash solution) has now been treated. The precipitate may contain iron and more chromium oxide, and the phosphates, etc., of the alkaline earths.

13. We will now proceed with a portion of this precipitate by first dissolving it in as small a quantity of hydrochloric acid as is possible, filter, and add to the solution (made as nearly neutral as possible) two or three drops of ferro-cyanide of potash (yellow prussiate of potash in solution); a blue precipitate is formed, proving the presence of *iron sesquioxide*. Wash another portion and fuse it in a small crucible with potassium nitrate (pure saltpetre) and sodium carbonate about equal parts. When cold digest with water; a yellow solution results, which produces a yellow precipitate with acetate of lead, showing the presence of *oxide of chromium*. This double finding of chromium oxide (for it was found before) is due to the relative quantity of iron present as related to chromium oxide present, which will not be entirely precipitated at one time in the presence of iron under these circumstances.

14. We now go back to the solution filtered off from the precipitate treated of in paragraph 11. This solution may contain zinc, manganese, nickel and cobalt. Digest with ammonium sulphide, wash the consequent precipitate and dissolve it in nitro-hydrochloric acid (aqua regia). It may be dis-

solved upon the filter by dropping the mixed acids and filtering through into a clean beaker, just as it could have been done in paragraph 11. This is convenient when the precipitate adheres too tightly to the filter to allow of scraping it off entirely. Digest this clear solution with potassium hydroxide (or caustic potassa) precisely as in paragraph 11. This potassa may be put into the beaker in small pieces of the stick, in which form potassium hydroxide generally is sold.

(*a*) The *solution* may contain zinc oxide.

(*b*) The *precipitate* may contain manganese, cobalt and nickel, as oxides. Pass hydrogen sulphide through the *solution* (*a*) until the precipitate (white zinc) has ceased to fall. Wash and agitate the precipitate (*b*) with a solution of carbonate of ammonia. The precipitate which now falls is the carbonate of *manganese*—confirm this by the blow-pipe. The solution from this last treatment may contain cobalt and nickel oxides. Evaporate it to dryness, redissolve in a few drops of hydrochloric acid, and again evaporate to a moist mass and divide the mass into two parts. Heat one portion with borax in the blow-pipe flame, a blue bead proves *cobalt*. Dissolve the other portion in water and add solution of cyanide of potassium slowly, a precipitate is formed which on continued adding of the potassium cyanide begins to redissolve. On adding hydrochloric acid it is again precipitated. It is *nickel*. Confirm with the blow-pipe.

15. In paragraph 9, paper *A* was put aside.

This paper contained the precipitate holding the copper of the ore *if any was present.* Digest this with ammonium sulphide (or potassium sulphide). A solution and a precipitate are formed. The *precipitate* may contain lead, mercury, bismuth, cadmium, besides copper, as sulphides. The *solution* may contain gold, platinum, antimony, arsenic, and tin as sulphides.

16. Treat the *precipitate* first, by boiling it with nitric acid. A black or browish residue remains undissolved. Take a hard glass tube, and having washed and dried the black residue, introduce some of it into the tube and heat it. It may act in three ways: (*a*) it sublimes without change; *mercury oxide* was present—test with blow-pipe; (*b*) it sublimes leaving a white powder which when moistened with ammonium sulphide turns black, proving it to be *lead sulphate;* (*c*) it sublimes, but as a mixture of *mercury sulphide* with minute globules of metallic *mercury*, showing that through some haste or lack of care, mercury as sub-oxide of mercury still remains when it should have been entirely precipitated as *chloride of mercury* at the first (paragraph 5).

17. We now proceed with the filtrate (obtained as stated in paragraph 16), from the black or brownish residue. Treat this with solution of carbonate of potash and wash the consequent precipitate, and then digest this precipitate in *cyanide of potassium* in excess, while it is moist. This may be done on the filter after changing the beaker, since this

filtrate or solution must be kept. The *insoluble* part may contain lead and bismuth as carbonates—the *solution* may contain copper and cadmium as double salts with cyanide of potassium.

18. Proceed with the *insoluble* part by boiling it with dilute hydrochloric acid. To one part of the resultant solution add sulphuric acid ; the precipitate indicates *lead*. To the other part, after concentration by evaporation, add a large quantity of water—a milkiness is produced indicating *bismuth*.

19. Into the *solution* (paragraph 17), after digesting with potassium cyanide, pass hydrogen sulphide —the *precipitate*, if formed, indicates *cadmium*—test it with the blow-pipe. To the solution add hydrochloric acid—*copper* sulphide will be precipitated; add a few drops nitric acid which will dissolve the copper sulphide, and then by adding ammonia in slight excess the solution has a blue color indicating *copper*.

20. We are now to treat the *solution* mentioned in paragraph 15. The insoluble part, paragraph 16, having been separated off as there stated, add to the solution acetic acid, and boil. If a precipitate be produced, collect a small portion, wash and heat it over a spirit-lamp upon a strip of platinum foil. If it burns with a bluish flame and leaves *no residue whatever*, it is *sulphur* and nothing more may be done—this part of the assay is exhausted. But if it leaves some residue, then several important elements may be present. Proceed, and to one part add a solution of chloride of tin (protochloride with a

drop of nitric acid added), a purple color is produced. To another part add a solution of proto-sulphate of iron—a brown precipitate is produced indicating *gold* in both cases.

To another part add ammonium chloride (solution), a yellow crystalline precipitate falls which marks *platinum*. *Arsenic* may be tested by the blow-pipe in the ore, but if the presence of sulphur, in larger quantity, prevents detecting a small quantity of arsenic, it may be detected thus: Take a part of the black or brownish precipitate resulting from the addition of acetic acid, and mix it with three times its bulk of nitrate of potash (saltpetre) and carbonate of soda. Project this mixture, a little at a time, into a Berlin crucible, in which a mixture of the same substances has been placed and is in fusion over a lamp. At conclusion, digest the fused mass with pure water; filter; add excess of nitric acid and heat; now add nitrate of silver; filter when cold, and add very dilute ammonia; a brown precipitation or coloring marks *arsenic*.

Dissolve another portion of the dark precipitate or residue from acetic acid in hydrochloric acid. Place in the solution a strip of metallic zinc—a pulverulent deposit takes place on the zinc, indicating *antimony*. If more proof be wanted remove the powder to a beaker and digest in nitric acid, when a white precipitate is formed. Digest it with a strong solution of tartaric acid, only a part may be dissolved, but filter; into the clear solution pass hydrogen sulphide and an orange-colored precipitate is formed, proving *antimony*.

In the last paragraph it was found that a part of the precipitate was not dissolved in the tartaric acid; dry it; place it on charcoal with a little cyanide of potassium and carbonate of soda, and turn the inner flame of the blow-pipe upon it; it is reduced to metallic *tin*.

In the above analysis provision has been made for the detection of sixteen elements. Of course, if no precipitates or signs appear at any one stage of the analysis, proceed immediately to the next, for it is not probable that any mineral will ever contain even one-half the elements mentioned in the assay, but the full number is given so as to reach any possible case.

II. DRY ASSAY OF ORES.

We have given the wet assay method, and we now give as much of the dry assay as may generally be called for.

What will be first needed in the dry assay are crucibles, scorifiers and cupels. CRUCIBLES for general purposes are made of coarse material, and are called Hessian. They are sold in nests of five or more. The only sizes of much value are those holding about 6 to 8 ounces. SCORIFIERS are flat, but thick, clay saucers intended to prepare the rough ore for the finer treatment by use of the cupel and in the assay furnace. The CUPEL is a little saucer of bone-ash, intended to be used on the floor or bottom of a heated muffle in the assay furnace. The MUFFLE is a clay oven of small dimensions, intended to protect the scorifier and cupel from the coals of the furnace. They can be obtained at any chemical warehouse.

An ASSAY FURNACE may be made of sheet-iron; it should be some 15 inches in diameter, with a grate near the bottom, and lined with either ordinary or fire brick.

In the accompanying figure is given the general form of one which has been used for years with perfect success.

A plain sheet-iron cylinder (Fig. 29) 18 inches high and 15 inches in diameter, with draft hole at A, muffle hole at B, and pipe-hole at C, and lined, as has been said, with brick, will answer all purposes of the best assays. The hole at C must have a collar and pipe either for a chimney or it must enter a chimney. B must be provided with a flanged door, as also the draft hole A. The top may have, loosely laid on, only a square sheet of heavy sheet-iron, and the whole placed upon a flat stone or a few bricks. Several heavy bars of iron nicked into the bricks will answer where there is no iron foundry at hand to cast a grating, D. Charcoal or coke may be used, or, where the draft is strong, a hard coal.

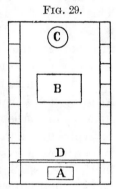

FIG. 29.

The crucible should be lined with charcoal finely pulverized and made pasty with molasses or any

syrup. This process is called "brasquing." Heat
the crucible before using, to dry out the syrup.

For field testing a small portable assay furnace,
using preferably some form of gaseous fuel, is of
great advantage. Such a furnace is made by E.
H. Sargent & Son, of Chicago, Illinois. It has the
advantages of only weighing 7 lbs., being about 5
by 8 inches, when set up is about 20 inches in
height, and it packs in a space of 1 cubic foot with
all the necessary material—the box then weighs
ready packed, some 25 lbs. (without mortar and
pestle); and lastly, one of its greatest recommenda-
tions is that refined petroleum is used as the fuel.
This form of fuel is much more easily obtained, and
is less dangerous than gasoline, which is the liquid
fuel most commonly used for assaying.

If the object is to obtain the amount of IRON IN
AN ORE, pulverize the ore to about forty to the inch,
weigh it, mix it with charcoal and cast the mixture
from a piece of paper on the bottom of the crucible,
cover it with charcoal an inch or two deep, drop in
two or three pieces of brick, and place the crucible
in the hottest part of the fire, cover all with coal
and gradually increase the heat and keep it nearly
at white heat for half an hour, draw it out, jar the
crucible down on a stone to settle the melted
button. When cool take out the contents, and the
metallic iron will be found with its slag attached.
Clean the button, weigh it, and the weight of the
ore used is to the weight of the button as 100 is to
the per cent. of iron in that ore ; that is, multiply

the weight of the button by 100 and divide by the weight of the ore used.

SCALES, WEIGHING, ETC. Any scales that weigh from ¼ oz. to ½ lb. or a greater amount will serve for the rough work in the field. The cheapest and lightest scale is one used for weighing letters, which weighs from ¼ oz. to 12 ozs.; but a better scale is a light spring balance, weighing up to 2 lbs., and divided into ½ and ¼ ozs.

The sample can best be weighed by laying it on a sheet of paper, turning up the edges, and tying them with a piece of string which can be hooked on to the scales.

For more delicate work, a small pair of scales weighing to $\frac{1}{100}$th of a grain is quite sufficient. Such scales may be bought at any chemical ware-house, made to pack and carry with ease and security. When in a fixed laboratory at home, the scales weighing to 0.0077 grain or half a milligram will save chemicals, time and work; but unless the analyst has an absolutely true average of the ton of ore most carefully chosen, the smaller the amount of ore used the more likely is the assay to prove deceptive when proportioned to the ton.

PULVERIZATION for the dry method should never be more than 50 or 60 to the inch. Smaller particles are apt to be lost or separated in the crucible. Obtain a piece of silk bolting cloth from a flour miller or from the source from which he gets his cloth, and select two or three grades, one for "wet analysis," which may be as fine as 80 to the inch.

Have a rim made by the tinner to tie on the sieving cloth, or use a cracked beaker glass, cutting it off by the method we have already given. (*See previous note, page* 67.)

GOLD AND SILVER ORES. These ores require preparation in the scorifier. Powder the ore, of which take about 50 grains ; of lead shavings take from 500 to 1000 grains, according to the probable amount of silver, much if much silver is supposed to be present, and of borax take about 50 grains. Mix the ore with half the lead and place it in the scorifier, spread the other half over the contents, and finally spread the borax over all. Put the scorifier in the muffle, close the door, and heat up to fusion—then the door must be partly opened, the heat increased, until the oxidized lead (litharge) covers the scorifier —take it out and pour the contents into an iron cavity or mould, separate the button and hammer it up into the shape of a cube. It is now ready for cupellation, as it contains all the gold and silver.

CUPELLATION. This process simply separates the lead from the gold and silver. This it does both by absorbing and by oxidizing. Cupels may be made, but they may be bought so cheaply that it is seldom worth the trouble to make them.

Push a cupel into the heated muffle, place the cube of lead in the cupel with little tongs, and heat up till the lead melts, watch the lead gradually wasting away until reduced to the size of the silver it contains, when the surface will become instantaneously bright and nothing remains but the silver containing the gold. Withdraw the cupel and cool and weigh the ball. The gold and silver must be separated by the wet process, thus: Dissolve the ball in strong nitric acid with heat till the acid boils; a dark powder precipitates; filter off the dark powder, it is the gold, and precipitate the silver by solution of common table salt or by hydrochloric acid. After all is precipitated drop into the white precipitate some pieces of zinc, add more hydrochloric acid—hydrogen gas is generated, which reduces the white silver chloride to powdered metallic silver. The gold and the silver may now be melted in separate crucibles, weighed and compared with the amount of ore used.

In these trials the lead should first be cupelled for its silver, and that subtracted from the silver found, as almost all leads contain some silver.

If it should be more convenient to melt the ore in a crucible rather than a scorifier, use the following flux: If the ore is composed chiefly of rock, pulverize, take 100 to 500 grains of ore, red lead 500 grains, charcoal powder 20 to 25 grains, carbonate of soda and borax together 500 grains—the more rock the more carbonate of soda, the more metallic bases the more borax. Place a little borax over all and melt till all is liquid, requiring about 20 minutes; withdraw, extract the button when cool, hammer up to a cube and cupel. Separate the gold and silver as before, but remember that the amount of silver must be three times that of the gold, and if there is reason to believe that there is not this

6

amount, some silver must be melted with the button, since the separation will not otherwise be complete.

T. S. G. Kirkpatrick recommends the following process of assaying gold quartz: Take 200 grains of ore, 500 of litharge, 6 of lamp-black and 500 of carbonate of soda; or, 200 grains of ore, 200 of red lead, 150 of carbonate of soda, 8 of charcoal and 6 of borax. Mix and put into a warmed crucible, and cover with half an inch of common salt. Fuse in a hot fire 30 minutes; cool and break the pot; clean the button with a small hammer.

If the quartz is very pyritous, take 100 grains and calcine "dead" without clotting, add 500 grains of red lead, 35 of charcoal, 400 of borax, and 400 of carbonate of soda, cover with salt and proceed as above. In each case cupel the button.

As the bone ash of which the cupel is made can absorb its own weight of metallic oxides, the cupel chosen should always exceed the weight of the button to be operated on, so as to have a margin. Boil the gold prill obtained from cupelling in nitric acid, which dissolves the silver and leaves the gold pure.

The above formulæ are open to modifications by the operator according to the apparent richness or poverty of the ore to be treated, and the presence and character of the basic impurities. In case there are oxides, a reducing agent is required; and if sulphides, an oxidizing agent. As a rule, employ a weight of litharge *twice* that of the ore, and of carbonate of soda the *same* as the ore. These reagents are added to control the size of the lead button, and to obtain one of suitable size for cupelling.

LEAD ORE, GALENA. The charge for the crucible is carbonate of soda, two or three times the weight of the ore, three or four tenpenny nails on top to absorb the sulphur, and a covering of salt or borax; heat to redness about 20 minutes. Pour the contents into a crucible and separate the button.

COPPER ORE. The wet assay is better than the dry, especially that by the burette, which we shall give later on under "Copper."

TIN ORE. If it is mixed with iron or copper pyrites it should be powdered and roasted, and then mixed with one-quarter of its weight of charcoal and subjected to great heat in a crucible for about 20 minutes. Jar it as in an iron assay, let it cool, and pick out the button or buttons, or pour it out while melted.

It may be reduced otherwise by melting the powdered ore with cyanide of potassium, 100 grains of ore to 600 grains of cyanide. Cool, extract button.

This ore is very hard and may be powdered to 60 to the inch.

MERCURY. These ores are easily reduced by simply heating and condensing the vapors in a cold bath as in using a retort and cool receiver.

ANTIMONY. Place about 2000 grains of ore powdered in a crucible having a hole chipped out in the bottom, and the hole stopped loosely with a piece of charcoal. Put this crucible into another half-way down. Then lute on the lid and put clay around the juncture of the two and put live coals around the upper crucible by placing some broken bricks around the lower on the grate, to keep the coals away from the upper. The antimony will melt and leave its gangue rock in the upper crucible while the lower will receive the melted metal.

BISMUTH, ZINC, MANGANESE, NICKEL, COBALT, and other metals should be reduced or analyzed by the "wet process" which we have already given. (In this chapter, V.)

An excellent fire lute is made of 8 parts of sharp sand, 2 parts of good clay, 1 part horse-dung; mix and temper like mortar.

CHAPTER VI.

SPECIAL MINERALOGY.

GOLD.

WE shall now proceed to a more definite and practical treatment of these two subjects, TECHNICAL MINERALOGY and ECONOMIC GEOLOGY, so far, only, as they may be of service in the work before us.

The first suggestion we have to make is that the best preparation for the general study of mineralogy is to gather a collection of the chief mineral substances with which the student is to come in contact. In many cases very small specimens are sufficient. As we proceed in our treatment of each substance it will occur to the reader what and how much he needs to obtain. But it should be emphasized that no amount of study on the part of the student, nor of description on the part of the instructor, can ever take the place of the actual specimen.*

Gold.—Gold is one of the most widely distributed metals, but generally speaking, accumulations of larger quantities of it are found only in a few localities. Traces of it pass from various ores into artificial products, for instance, into litharge, minium,

* For list of specimens, see end of book.

(85)

white lead, silver and copper and coins made therefrom, etc. Minute quantities of gold (about 13 grains in 1 ton) have been found even in sea water as well as in clay deposits.

The chief supplies of gold are at the present time obtained from the United States (California, Nevada, Arizona, Montana, Utah, Alaska, Colorado) from British Columbia, Nova Scotia, Mexico, Peru and Brazil, from Australia (especially Victoria, New South Wales, and Queensland), Tasmania, New Zealand, and in Africa (Natal, the Transvaal, etc.). The Ural Mountains and Siberia also yield considerable gold. In Europe only Transylvania and Hungary are of any importance.

Gold occurs almost exclusively in the metallic state, either *in situ*, in quartz rock, especially along with quartz, pyrites and hydroferrite; also as gold sand, in dust or grains, leaflets and rounded pieces (nuggets), in the sands of rivers or in alluvial soils, consisting chiefly of clay and quartz sand along with mica, water-worn fragments of syenite, chlorite slate, grains of chrome iron and magnetic iron, spinel, garnet, etc. In the metallic state it contains always more or less silver as electrum. According to recent analyses native gold contains:

	Transyl- vania.	South America.	Siberia.	California.	Australia.	
Gold	64.77	38.14	86.50	90.00	99.2 and	95.7
Silver	35.23	11.96	13.20	10.06	0.43 "	3.8
Iron and other metals.	—	—	0.30	0.34	0.28 "	0.2

Siberian, Californian and Australian gold contain not unfrequently osmiridium, palladium and platinum. Mexican rhodium-gold contains 34 to 43 per cent. rhodium. Gold amalgam is found in California and Columbia. The so-called *black gold* which occurs in nuggets in Arizona and at Maldon, Victoria, in granite and quartz lodes, is crystalline and silver-like when freshly fractured, but soon turns black in the air. It is *bismuth-gold*, with 64.211 gold, 34.398 bismuth and 1.591 gangue. Gold is also often met with in native tellurium and silver telluride, sometimes in iron pyrites, copper pyrites, in blende, in arsenical pyrites, and galena. To detect a content of native gold in pyrites bring a few drops of mercury into a porcelain crucible, put a perforated piece of cardboard in the crucible so that it rests a short distance above the mercury, place a small package of pyrites over the hole in the cardboard, heat the crucible for some time and watch with the pocket lens for the appearance of white stains of gold amalgam, which on rubbing with a brush or feather becomes lustrous.

Gold crystallizes in the isometric system, but crystals are seldom found. Figs. 30 and 31 represent gold crystals. Twin crystals are also occasionally found. In Sonora, California, Blake found gold in hexagonal prisms. Fig. 32 shows the finest gold dust 700 times magnified, and Fig. 33 a reduced illustration of a lump of gold which was found at Forest Creek, Victoria, Australia. It weighed more than 30 pounds, and was 11.33 inches long and 5.15 inches wide. The largest nugget of gold ever found was at Ballarat, Australia. It weighed over 191 pounds, and was 20 inches long and 9 inches wide.

The specific gravity of gold is 16 to 19.5, accord-

FIG. 30.   FIG. 31.   FIG. 32.

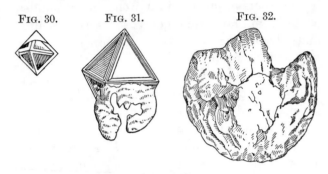

ing to the amount of alloy ; hardness 2.5 to 3.0.   It is the only *yellow, malleable* mineral found in the natural state.   Its color varies from pale to deep yellow.   In some localities, such as in New South

FIG. 33.

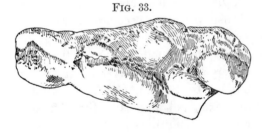

Wales, Australia, and Costa Rica, it is often found of a very light color, but it presents the same color from whatever direction it is looked at, and to the prospector this is a guiding test.   Indeed one of the

most important and useful accomplishments for
gold exploitation is "an eye for color." Native
gold possesses a peculiar color which is readily
recognized, although the gold may be alloyed with
silver or copper, and its color will in an instant distinguish it in the eye of the expert from any condition of pyrites, whether iron or copper pyrites.

Gold grains will always flatten when struck with
a hammer or between two stones, whereas other
minerals similar in color will break into fragments.
Or if the doubtful particle is coarse enough, take a
needle and stick the point into the questionable
specimen. If gold, the steel point will readily prick
it; if pyrites or yellow mica, the point will glance
off or only scratch it.

Under the blow-pipe, on a piece of charcoal, gold
may melt, but on cooling it always retains its color;
any other mineral will lose color, become blackened, or will be attracted to the end of your pen-
knife blade, if that blade has been previously
magnetized, and the unknown substance contains
iron.

Gold imparts no color to boiling nitric acid. It
will not dissolve in nitric or hydrochloric acid
separately, but it does dissolve in the two when
combined, and then the acid is known as nitro-
muriatic acid or aqua regia. Proportions: one
nitric to four muriatic.

But it is not always a trustworthy sign that particles are gold because they will not dissolve in
nitric acid. Some seemingly gold-colored particles

will not dissolve in nitric acid, and yet contain not a trace of gold.

The simplest instrument for the discovery of gold and the estimation of the value of an auriferous material in which the gold is contained in a free state, is the ordinary miner's pan, a circular dish of Russian sheet-iron, about 12 inches wide and 3 inches deep, with sloping sides. There should be a slight indentation all round where the sides join the bottom, so as to afford lodging for the gold grains, and the more rusty it is the better. A frying pan free from grease will answer very well on a pinch. The South American *batea*, Fig. 34, made

FIG. 34.

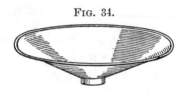

of hard wood in a solid piece, and hollowed out like a shallow funnel, is a superior implement when in capable hands. Another good substitute for this pan is a kind of magnified shovel without handle made of linden wood and provided with a vertical wall on three sides. The wooden implements should be slightly charred on the surface to show up the gold grains, and should not have been used to hold mercury or amalgam.

The object of *panning out*, as the operation with the pan or batea is called, is to settle and collect at the bottom of the pan the heaviest portions of the material subjected to the test. This is effected by filling the pan to not much more than half its content with the material to be tried, then provide a hole full of " still " water, which hole must be large and deep enough to allow of the free swinging of the pan under the surface of the water. Now dip the pan with the material into the water, quite filling it, rub the stones and gravel until the clay

FIG. 35.

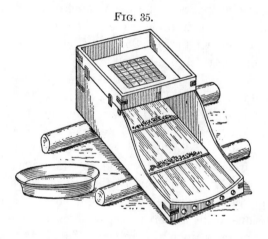

that may be adhering to them is removed and dissolved, then give the pan a few vigorous swinging rotary shakes, let the water run off, shaking the pan from right to left and left to right all the time. When the water has run off, the material should be lodged near the lip of the pan, which, being dipped into the water and raised above it, allows the flow of the water to carry off the lighter portions of the sand and gravel. After several such dips the pan is shaken as before, and the whole process repeated until no more than about a dessert-spoonful is left. This is then carefully examined with a pocket-lens for any mineral or metal it may contain. The gold will be immediately recognized by "an eye for color." Where water can be had, a pan is the most efficient instrument a man can travel with in his gold-seeking journeys.

A crude apparatus formerly much used in California and Australia is called the *cradle* or *rocker*. This, as shown in Fig. 35, is a trough of some 7 feet in length and 2 broad. Across the bottom of this several bars are nailed at equal distances, and at the upper end a kind of sieve is fixed at about a foot above the bottom. This whole arrangement is mounted upon rollers. To operate the apparatus four men are required. One man digs out the earth from the hole, a second supplies the cradle sieve with this auriferous earth, a third keeps up a supply of water which he pours upon the earth in the sieve, while a fourth keeps the machine continually moving upon the rollers. The large stones washed out are removed by hand from the sieve, and the water at the same time washes the smaller substance through, which is slowly carried towards the lower end of the trough by a slight inclination given to the whole. Thus the flow of water tends to keep the earthy particles in suspension so as to allow of their washing off, while the heavier portions of gold are obstructed in their flow, and retained against the cross bars fixed to the cradle bottom. These are removed from time to time and dried in the sun, when, after blowing away lighter particles, the metal only further requires to be melted.

A more efficient apparatus is the *long tom*, Fig. 36. This is a trough about 12 feet in length by 20 inches in width at the upper end, and widening to 30 inches at the lower end. It is about 9 inches deep and has a fall of 1 inch to a foot. An iron screen is placed at the lower end (cut off in the manner shown in the illustration) where large

FIG. 36.

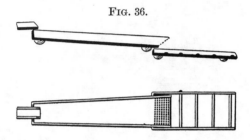

stones are caught, and below this screen is the riffle box, 12 feet long, 3 feet wide and having the same inclination as the upper trough. It is fitted with several riffles, in which mercury is sometimes placed. Much more work can be done with this appliance than with the cradle, which it has generally superseded. Of course the gold must be coarse and water plentiful.

Washing the gold dirt is also affected by *sluices* having an inclination of about 8 feet in 12 feet. These sluices consist of a series of troughs formed by planks nailed together, the length of each being about 10 or 12 feet, the height 8 inches to 2 feet, the width 1 to 4 feet. By making one end of the bottom plank of each trough 4 inches narrower than at the other, the troughs can be telescoped into one another and so a sluice of very great length can be formed. Across the inside of the bottom-planks, small narrow strips of wood 2 inches or so thick, and 3 or more inches wide, are fixed across, or some-

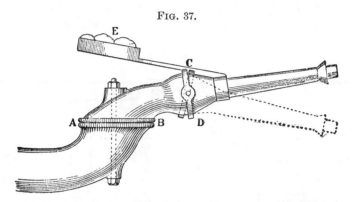

FIG. 37.

times at angles of 45° to the side of the trough, at short intervals apart. Running water washes downward the earth thrown into the sluice, which is open on the top side, and the gold dust accumulates, sometimes assisted by the aid of mercury allowed to trickle out of a vessel from riffle to riffle, in front of the bars, while the lighter matter is washed downwards.

A still more effective method is what is called *hydraulic mining*, and under favorable circum-

stances, such as a plentiful supply of water with
good fall and extensive loose auriferous deposits, a
very small amount of gold to the ton can be made
to give paying returns. The water is conducted in
flumes or pipes to a point near where it is required,

FIG. 38.

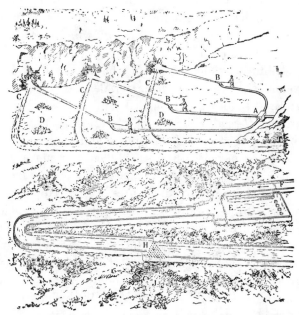

thence in wrought-iron pipes gradually reduced in
size and ending in a great nozzle somewhat like
that of a fireman's hose. Figs. 37 and 38 show the
arrangement. Fig. 37 exhibits the mouth-piece
movable at *A B* in an ascending, and at *C D* in an
inclined direction. *E* is a lever loaded with

weights, which facilitates the adjustment of the
mouth-piece by the operator in any direction. The
method of operating the arrangement will be seen
from Fig. 38. *A* is the water-distributor, *B* the
nozzle, *C* channels for carrying off the debris detached from the ledge; *D*, piles of larger pieces of
rock which are finally comminuted. *T* is a tunnel
through which the water reaches the gutter, provided with the grating *F* through which the finer
stuff falls into the shallow settling basin *E*, and is
distributed by blocks *G*, while the principal mass of
water with the coarser material passes over the
grating *F* into the principal sluice in which the
grating *H* retains the larger pieces which are then
thrown out at *J*. The basins *E* and the principal
sluice are paved with wooden blocks or stones between which mercury is placed. The amalgam
formed is freed from admixtures in a mercury bath,
pressed through sail-cloth, boiled in sulphuric acid
and distilled.

Burning and Drifting. The labor of removing the
barren gravel which overlies the pay dirt is very
great, but ordinarily this is undertaken when the
thickness is not considerable. With increasing
thickness a point is soon reached where the task of
removing it becomes so formidable that the miner
will not make the attempt unless he believes that
there is rich pay dirt beneath. In this event the
practice is adopted of sinking shafts through the
barren material to the pay dirt, and extracting the
pay dirt by means of tunnels or drifts along the sur-

face of the bed-rock. This method of working has been adopted only lately, but promises to be very important. The ordinary methods of sinking, drifting, timbering, stoping, etc., have been peculiarly modified in the Forty-mile District, Alaska, on account of the exceptional character of the climate, and these modifications have spread from this district over the rest of the gold diggings. Owing to the severity and length of the winters the gravels are frozen during seven or eight months of the year. The miner who desires to sink a shaft waits until the cold season arrives, and then sinks through the frozen ground, which is so firm that the shafts or drifts do not need timbering for the sake of support. In sinking or drifting, instead of employing powder and pick, as elsewhere, a small fire is built at the bottom of the shaft which is being sunk, or at the face of the drift which is being run, and thus the gravels are thawed out for some distance and can be easily taken up and brought to the surface. It takes a surprisingly small amount of wood to run a drift through the frozen gravel for a long distance. In this way the pay dirt is extracted and accumulates on the surface until spring, when it is shoveled into sluices and the gold is separated by washing, panning, blowing and amalgamation in the manner previously described. One large chamber or "stope" thus excavated in the gravels of Miller Creek in the Forty-mile District, is said to have measured 64 by 32 feet, and 19 feet in height, with only 8 feet of barren gravel between it and the surface ; and yet

7

this stood firmly until spring, when the gravels thawed and the stope caved in.

For *lode prospecting* a pestle and mortar should be carried. The handiest for traveling is a mortar made from a mercury bottle cut in half, and a not too heavy wrought-iron pestle with a hardened face. To get the stuff to regulated fineness a fine screen is required, and the best for the prospector who is often on the move, is made from a piece of cheese cloth stretched over a small hoop. It is often desirable to heat the rock before crushing, as it is thus more easily triturated and will reveal all its gold. Having crushed the gangue to a fine powder, proceed to pan it off in the same manner as washing out alluvial earth, except that in prospecting quartz one has to be much more particular, as the gold is usually finer. Take the pan in both hands and admit enough water to cover the pulverized substance by a few inches. The whole is then swirled around and the dirty water poured off from time to time till the residue is clean quartz sand and heavy metal. Then the pan is gently tipped and a side to side motion given to it, thus causing the heavier contents to settle down in the corner. Next the water is carefully lapped in over the side, the pan being now tilted at a greater angle until the lighter particles are all washed away. The pan is then once more righted and very little water is a few times passed over the pinch of heavy mineral, when the gold will be revealed in a streak along the bottom. In this operation, as in all others, only practice will make perfect, and a few practical lessons are worth whole pages of written instruction.

J. C. F. Johnson * gives the following directions for making an amalgamating assay that will prove the amount of gold which can be got from a ton of a lode. Take a number of samples from different parts, both length and breadth. The drillings from the blasting bore-holes collected make the best test. When finely triturated weigh off one or two pounds, place in a black iron pan (it must not be tinned) with 4 ounces of mercury, 4 ounces common salt, 4

FIG. 39.

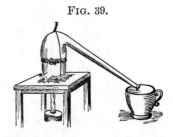

ounces soda, and about half a gallon of boiling water. Then with a stick, stir the pulp constantly, occasionally swirling the dish as in panning off, till you feel certain that every particle of the gangue has come in contact with the mercury. Then carefully pan off into another dish so as to lose no mercury. Having got your amalgam clean, squeeze it through a piece of chamois leather, though a good quality of new calico previously wetted will do as well. The resulting pill of hard amalgam can then be

* "Getting Gold." London, 1897.

wrapped in a piece of brown paper, placed on an old shovel, and the mercury driven off over a hot fire. Or a clay tobacco pipe, the mouth being stopped with clay, makes a good retort. To make such a retort, Fig. 39, take two new tobacco pipes similar in shape, with the biggest bowls and longest stems procurable. Break off the stem of one close to the bowl and fill the hole with well-worked clay. Set the stemless pipe on end in a clay bed, and fill with amalgam, pass a bit of thin iron or copper wire beneath it, and bend the end of the wire upwards. Now fit the whole pipe, bowl inverted, on to the under one, luting the edges well with clay. Twist the wire over the top with a pair of nippers till the two bowls are fitted closely together, and you have a retort that will stand any heat necessary to thoroughly distill mercury. The residue, after the mercury has been driven off, will be retorted gold, which, on being weighed and the result multiplied by 2240 for 1 pound assay, or by 1120 for two pounds, will give the amount of gold per ton which an ordinary battery might be expected to save. Thus 1 grain to the pound, 2240 pounds to the ton, would show that the stuff contained 4 ounces, 13 pennyweights, 8 grains per ton.

Although not strictly within the scope of this small book, the process of extracting gold from lode stuff and tailings by means of cyanide of potassium, which is now largely used, may be thus briefly described : It is chiefly applied to tailings, that is, crushed ore that has already passed over the amalgamating and blanket tables. The tailings are placed in vats, and subjected to the action of solutions of cyanide of potassium of varying strengths down to 0.2 per cent. These dissolve the gold, which is leached from the tailings, passed through boxes in which it is precipitated either by means of zinc shavings, electricity, or other precipitant. The solution is made up to its former strength and passed again through fresh tailings. When the tailings contain a quantity of decomposed pyrites, partly oxidized, the acidity caused by the free sulphuric acid requires to be neutralized by an alkali, caustic soda being usually employed.

When "cleaning up," the cyanide solution in the zinc precipitating boxes is replaced by clean water. After careful washing in the box, to cause all pure gold and zinc to fall to the bottom, the zinc shavings are taken out. The precipitates are then collected, and after calcination in a special furnace for the purpose of oxidizing the zinc, are smelted in the usual manner.

OTHER FORMS AND CONDITIONS. Beside in the condition of simple native gold, this metal is found, as previously mentioned, in intimate mixture with pyrite (iron sulphide). It does not seem to be a compound, but, as we have said, a mixture or minute association. This seems evident from the fact that when the sulphur is removed from the pyrite and the iron rusts down, the gold particles appear with their own color and characteristics in cavities of various rocks, which, when crushed or water-worn, release the particles or pieces to be washed down and mingled with sands and gravels of lower levels, or perhaps the beds and channels of rivers. This is "placer gold." Where gold has not yet been thus released, it is found in association with iron, and especially with quartz in veins. In some instances the gold in quartz is disseminated in particles so exceedingly fine as to require the lens to reveal it.

Nevertheless quartz is not the only mineral which contains gold, although it is the world's great paying source of gold. Some of the other minerals contain it. It is found in yellowish-white, four-sided prisms, and in small white grains as large as a pea, and easily crumbles. In this condition the gold is amalgamated with quicksilver in the proportion of 38 gold to 57 quicksilver, and is known as "gold amalgam." It is very easily tested by heating upon a piece of charcoal by a blow-pipe, when the quicksilver volatilizes and the gold remains.

Gold in paying quantities is found in numerous combinations, and must be discovered and extracted either chemically, by the "wet method," or by assaying in the crucible by means of the cupel and furnace, when it cannot be separated on the spot by the blow-pipe. These methods are taught in any book upon the assay of gold.

Geology of gold. In studying the geological aspect of this subject and making the practical application of our knowledge to the search, we may state that the original position of gold must have been in great depths. From these depths it has been brought up by the upheaval of the granitic rocks, and perhaps, along with basaltic and other intrusions shot up from immense depths. In the course of ages the attrition and breaking down of these higher or uplifted levels, and the long-continued floods, rains and the waves of ancient oceans and other disintegrating forces which produced the sedimentary rocks, at the same time liberated the gold which was incapable of decomposition. The gold thus found new and varied resting places in the sedimentary rocks of various ages, and in all the conditions which the surface might assume.

The quartz rocks are neither igneous nor sedimentary, but are supposed to have been in liquid form as solutions of silex, which, during long periods of time, gradually deposited the silex and whatever they contained, the water disappearing by evaporation or absorption.

Frequently, cellular quartz has been found with gold within the cells, the material which surrounded the gold having become decomposed, and, thus releasing the undecomposed gold, the latter is found in the cells of the quartz.

Gold, therefore, is to be expected and looked for in granitic regions (Fig. 40), and in those rocks and from those gravels and sands which owe their origin to such regions. It requires much judgment, general exploration, and knowledge of the region before the prospector can, with probability, expect to meet with gold, or before he should begin the search.

But with a full knowledge of the geologic condition of the country, and acting in accordance with the above facts, the prospector will soon come upon traces of gold, if any exist.

In looking for indications, the prospector should never pass an ironstone " blow-out " without examination, as, according to the German aphorism, " the iron hat covers the golden head," or as the

FIG. 40.

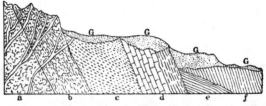

Section showing the two conditions under which gold is usually found in rock and drift.

THE STRUCTURE OF THE URAL MOUNTAINS.—*a.* Granitic and gneiss rocks penetrated with greenstones and porphyrytic rocks containing gold finely disseminated. *b.* Micaceous, talcose, and argillaceous slaty rocks, supposed to be Laurentian and Cambrian. *c.* Silurian and Devonian strata. *d.* Carboniferous, limestone and grits. *e* Coal measures. *f.* Permian and newer rocks. *G, G, G, G.* Drift, filling hollows in rocks with gold, especially at the base of the drift.

Cornishman puts it, "iron rides a good horse." The ironstone outcrop may cover a gold, silver, copper or tin lode.

Besides the general instructions given above, considerable study should be devoted to the peculiar and seemingly irregular deposits of gold where it does not appear to have been washed down from any higher levels.   For instance, in California and some other districts free gold has been found in drifts and sand and in the beds of streams which have not only been filled up, but have been buried under regions of sandstone or other rocks, but the whole country has apparently been raised, or the surrounding region has sunk so as not to show any very considerable elevation beyond where the gold deposits have been formed. But, even in this case, the general rule has been shown to be correct, for these deposits have been proved to be in the beds or channels of ancient rivers, which had either been dried up and overflowed by vast eruptions of lava or basalt, and again by floods bringing new soil and creating sedimentary rock, or the country has been raised, or subsidence of a great extent of land has taken place. In many cases, however, no subsidence has occurred, but only overflow and filling up through ages, and the actual sources still remain elevated.

Such events as we have just described do not transpire without leaving, in some parts, traces or features or material, which, to the practiced eye of a skillful prospector, are evidences of some such movements and changes, and he may proceed to make a successful opening only after he has carefully examined a large tract of country, for it is from extended survey that he may the more wisely judge of the relation of superficial parts to the greater depths of even small areas.

Those rocks which lie more immediately over the granite, and which, although they owe their origin to a sedimentary condition, have been subjected to heat and heated waters, as is supposed, we have called "metamorphic rocks." ' But they have been, probably, first formed from the disintegration of the most ancient rocks, and have brought with them fragments of gold. These metamorphic rocks have been changed from ordinary sedimentary rock by the action of heat and by pressure, and the influence of such treatment may be suspected by their appearance as crystalline in their composition ; that is, the fine grains which compose them, as well as the larger grains, are angular, whereas the materials of purely sedimentary rocks are fine without angular shape. The larger part of granite is supposed to have been metamorphic or changed, as the word means, or "altered" merely by the action of heat into a crystalline form or mass.

The igneous rocks are those whose forms are due to having been melted and driven to the surface through fissures in the overlying rocks. They are variously composed of feldspar, hornblende, little quartz, with comparatively small proportions of other substances, and are called by various names according to the composition. The metamorphic granite contains quartz, feldspar, and mica ; the igneous granite contains little or no quartz. Syenite-granite contains hornblende in place of mica. Sometimes the mica is very black, as hornblende is, and in that case may be distinguished from the latter by its more easy cleavage, as we have shown, under a sharp pen-knife ; this black mica is the kind we have described as *biotite* (p. 18). There is a syenite which contains no quartz, called hyposyenite. These rocks are not the original home of gold, but at present it is very largely in these metamorphic rocks that the most paying gold is to be found, more especially in the quartz veins which have intersected these rocks. One, therefore, of the most important studies of the prospector is to acquaint himself familiarly with the appearance, the locations, and the departures of these metamorphic rocks. In many places where the alluvial gold, derived from the gold-bearing gravels, has almost ceased to be worth working, there still remain sources undiscovered, and these sources may probably be traced back even yet to some out-crop or to some ancient elevation now having subsided.

The above remarks are applicable to explorations for other metallic ores than gold. They apply to silver, and especially to tin ores, and with some modifications to copper ores and to quicksilver, as we shall show.

GOLD IN COMBINATION. We have been speaking of gold as native and alone. But it must not be thought that this condition is the only one in which paying gold is found. The combinations of gold with various oxides and sulphides of other metals are very valuable, and should be studied.

In almost all gold-bearing regions the iron sulphides carry much gold, and in some regions the paying gold is found only in this substance. Hence, it is well for the prospector to determine the presence of gold in the pyrite or whatever sulphide may present itself. We, therefore, state a method or two of determining the fact that gold exists in this substance.

1. *To separate gold in metallic sulphides, for instance, iron pyrites.* Powder the sulphide as finely as possible. Put about an ounce into a Hessian crucible and heat to a very low red heat for an hour, or until there is very little escape of sulphur fumes. Remove the crucible and put its contents into a porcelain dish. Pour over the roasted powder three fluidounces of strong nitric acid, by drops, until all violent action ceases. Add water, 8 or 10 fluidounces; the gold, if any, will appear as a very fine black powder; filter and dry, pick out a small particle of the powder and mash it upon a hard surface, iron or agate, in an agate mortar; if it is gold, it will show the gold color. A sufficient quantity of the dried powder may be placed upon a piece of charcoal, and by means of either *O* or *I* flame of the blow-pipe it may be melted, and both by its color and softness be proved to be gold.

There is a difficulty in this process which the prospector may not be able easily to overcome, and that is the necessity of using the strongest nitric acid. If he has a little laboratory he may readily make his own nitric acid of sufficient power, and then he possesses the simplest and quickest method of treating sulphides or any gold-bearing pyrites. The process is as follows: This acid may be made from common saltpetre and sulphuric acid of commerce. Dry the saltpetre after breaking it into small lumps

of a half inch in diameter, carefully drop the lumps
into a glass retort, hang the retort on a wire or
stand, and introduce the beak into a glass bottle.
Place the bottle in a basin of cold water and you
may now apply the heat of a lamp, keeping the
flame low and five or six inches off from the bottom
of the retort. A coal-oil lamp with a short chimney
may be used, and the heat regulated to a point at
which brownish vapors appear in the retort. Keep
enough acid in the retort to barely cover the saltpetre, and keep cool water in the basin, and the
vapors come over and condense without much
trouble.

Stop the operation when the vapors cease to come
over, and the mass in the retort seems to settle down
to an even surface. Then draw out the beak of the
retort and put the glass stopper into the bottle, and
keep the bottle away from light and heat. Wash
out the retort, and if you require more nitric acid
renew the operation. The retort should be tubu-
lated to allow of adding sulphuric acid during the
operation if needed.

This acid is a yellowish-brown liquid and is
known as "fuming nitric acid," and is one of those
very active and convenient aids in the laboratory
which cannot readily be purchased, and, therefore,
must generally be made; but so little of it may be
used that a small quantity goes a great way, and it
will effect a result which the strongest and purest
chemically-pure nitric acid fails to produce. Its
effect is to release the gold from the combination ot

iron and sulphur by oxidizing the latter as well as the former, and rendering them soluble in water, while the gold remains in metallic form of an exceedingly fine black powder, as has been said.

2. Another method of detecting and separating the gold, where the above one cannot be used, is by pulverizing the sulphide ore very finely and mixing it with three or four times its weight of caustic potash or caustic soda, and then subjecting the crucible, which contains the mixture, to a low red heat till all the contents cease agitation and become perfectly tranquil. Then remove the crucible, wait till all is cool, and then add hydrochloric (muriatic) acid in an amount equal to three or four times the bulk of the mass. To this, after standing three or four hours in a warm place, add the usual nitric acid (about an ounce), after transferring all the liquid to a porcelain dish, or, next best, to a beaker-glass. Let it stand in a warm place for about an hour, then add a little more nitric acid (about half ounce), stir it well with a glass rod or strip of glass, and let it stand again for an hour or two. Examine carefully, and if it seems to have been dissolved more thoroughly than before, add a little more nitric acid and warm again, stirring well as before. If no more seems to be dissolved, then filter and wash the sediment in the filter and let it dry, and remove the filter and contents for further examination. Now precipitate the gold from the filtrate by pouring into it a solution of ferrous sulphate. [Any clear green crystals of " copperas " (sulphate of iron)

of the drug store, filtered, after saturated solution
in clean rain-water and kept in corked bottles, will
answer this purpose.] Let the solution stand in a
warm place for an hour, drop in a few more drops,
and if any further precipitation takes place, add
half an ounce of the sulphate, stir it again, let it
remain an hour longer in a warm place till all precipitation ceases. Decant the supernatant clear
water and transfer the remainder to a filter-paper
carefully, and a little at a time, to avoid breaking
the filter-paper, then rinse the porcelain dish to get
all particles upon the filter-paper, and when all the
liquid has passed through, let it dry, and remove
all the contents of the paper to a small porcelain
capsule or crucible, and apply the heat of the blow-
pipe to burn off the paper or any organic substance
which may have got into the powder; the gold
remains, which may be gathered upon charcoal and
melted into a globule by the concentrated flame of
the blow-pipe, if in small quantity. Lastly, examine the contents of the filter which was laid
aside; and, if any appearance of gold is noted,
separate it under examination by a pocket
lens.

The high value of gold renders even a grain of
gold to the pound of ore, if that pound is an average pound in the ton, worth $80 to the ton of 2000
pounds. Hence, a pyrites which contains a half
grain to the half pound may prove too valuable to
neglect. In the Brazils, in deep mines, the ore
yields only half an ounce to the ton of ore, and yet

it is mined at a profit.*   In California, a continuous yield of three-eighths to half an ounce of gold to the ton of quartz is considered profitable working.†

It must be remembered, however, that the above process of extracting the gold from a pyritous ore does not extract with perfect accuracy all the gold unless conducted with more care and time than we have suggested, but it is sufficient to reveal the fact that the ore is valuable.

3. The following method requires more time and care and the use of a little furnace, but will give very accurate results.   Pulverize the ore supposed to contain any gold, whether pyritous or not.   Heat it in a crucible very gradually at first, increasing the heat to drive off as much sulphur as possible, frequently stirring it and increasing the heat till all fumes seem to have escaped.   Withdraw it and prepare a crucible (clay or Hessian crucible), by dipping it in a strong solution of borax in water, and heating the crucible and repeating the dipping and heating till the crucible shows a glazed inside. Then transfer all the roasted powdered ore, after weighing it (if you desire relative quantity), into the crucible, and cover it with the following mixture (called a flux): Six times the weight of ore in litharge, one of dry borax, and about twenty grains of charcoal pulverized.   Heat slowly at first, not allowing much foaming, until all is quiet and the metal button settles down at the bottom of the cru-

* Makins' Metallurgy, p. 227.
† Davies' Metalliferous Minerals and Mining, p. 64.

cible. Cool and break the crucible to extract the button of metal, which is now ready for cupelling. (*For this process, see p. 80.*)

We have given these three methods of separating gold from all the usual ores, any one of which may readily be used, and a little practice will enable the operator to be expert in their use. A great deal more depends upon the skill of the operator than upon the cost of his appliances.

It has not been thought necessary to give a list of places in the world where gold has been found, but in view of the excitement created by the rich finds of gold, in July, 1897, in the Klondike district, Alaska, it may be of interest here to give a brief description of the Yukon gold district, which besides the Klondike, comprises the Hootalinqua, Stewart, MacMillan, Forty-Mile, Sixty-Mile, Birch Creek, Munook Creek, Tanana and Koyukuk districts.

Throughout nearly the whole of Alaska gold is found disseminated in the detritus which has been derived from the abrasion of the solid rocks. Often it is in such small amounts that it cannot be profitably extracted, but sometimes it is concentrated by water action in such a degree as to invite mining. Thus far the profitable deposits have all been found in or near the beds of the present streams. These recent gravels may be divided into two chief classes. In the larger streams accumulations of gravel are made in places of slackening current, such as the inner or concave sides of curves. These accumulations are called bars, and often contain

8

much gold. The other occurrence is in the small gulches which feed the larger streams. In the bottom of these gulches the gravels are frequently very rich in gold, and as these are easily worked, they constitute at the present time the most important class of placer deposits.

The gold of the Yukon district is chiefly derived from quartz veins, which are found most abundantly in the schists of the Forty-Mile and the Birch Creek series, although not infrequently in the igneous and pyroclastic rocks of the Rampart series. It is also derived, although to a far less extent, from impregnated shear zones, which occur especially in the Rampart series. Of the quartz veins one set is sheared and one unsheared. The first is difficult to follow, for the veins are broken and non-persistent. The veins of the second set are often persistent and wide, and in some cases may be mined profitably. Impregnations along shear zones may also in some cases be sufficiently rich in metallic minerals to form ores under favorable conditions; and the rock in the region of these shear zones is often unfaulted, so that these ore bodies may be expected to be comparatively persistent.

The quartz veins are connected with dikes, chiefly light-colored crystalline rocks such as granite and aplite. This should be kept in mind in prospecting, and auriferous veins may be looked for in the schists near the dikes. In some cases, although not so commonly, they may also occur at some distance from a dike.

These gold-bearing rocks form a definite belt, extending in a general way from the Lower Ramparts of the Yukon and below to Dease Lake and other mining districts in British Columbia, a distance in a straight line of about a thousand miles. Of this distance, 400 or 500 miles is in American territory. The width of the belt varies chiefly with the minor folding, which has accompanied the greater plications. In this belt not only the gold-bearing veins, but the richest placers are found. This is naturally the case, since the gold in these placers is worn out of the solid rocks. It is especially true that the rich gulch gravels are in this belt, and also the most paying bar gravels, although fine gold in some cases may be carried somewhat outside the belt, and may be sufficiently concentrated in favorable situations to pay for washing.

The Birch Creek, the Forty Mile and the Klondike districts are all in this belt, and are all in the schistose rocks, and in these rocks new deposits of value may be looked for. Some placer diggings of value may also be found in the rocks of the Rampart series, but as a rule higher horizons are probably barren, save in exceptional cases. Conglomerate made up of the detritus from the schistose Birch Creek and Forty Mile rocks should be prospected, however, since they may prove to be fossil placers. Ancient gravels lying above the present stream channels should also be kept in mind, for they may in places contain sufficient gold to be profitably mined.

Rule for ascertaining the amount of gold in a lump of auriferous quartz, according to Phillips:

The specific gravity of gold is 19.000.

The specific gravity of quartz is 2.600.

These numbers are given here merely for convenience in explaining the rule; they do not accurately represent the specific gravities of all quartz and quartz gold. (The quartz gold of California has not, on an average, a specific gravity of more than 18.600.)

1. Ascertain the specific gravity of the lump. Suppose it to be 8.067.

2. Deduct the specific gravity of the lump from the specific gravity of the gold; the difference is the ratio of the quartz by volume: 19.000—8.067 = 10.933.

3. Deduct the specific gravity of the quartz from the specific gravity of the lump; the difference is the ratio of the gold by volume: 8.067—2.600 = 5.467.

4. Add these ratios together and proceed by the rule of proportion. The product is the percentage of gold by bulk: 10.933 + 5.467 = 16.400. Then, as 16.400 is to 5.467, so is 100 to 33.35.

5. Multiply the percentage of gold in bulk by its specific gravity. The product is the ratio of the gold in the lump by weight: 33.35 × 19.00 = 643.65.

6. Multiply the percentage of quartz by bulk (which must be 66.65, since that of gold is 33.35) by its specific gravity. The product is the ratio

of the quartz in the lump by weight: 66.65 × 2.60 = 173.29.

7. To find the percentage, add these two ratios together and proceed by the rule of proportion: 633.65 + 173.29 = 806.94. Then as 806.94 is to 633.65, so is 100 to 78.53. Hence, a lump of auriferous quartz having a specific gravity of 8.067, contains 78.53 per cent. of gold by weight. (The Mines, Miners, and Mining Interests of the United States in 1882, by Wm. Ralston Balch, Phila., p. 761.)

CHAPTER VII.

TELLURIUM MINERALS. Tellurium is the only metal which has hitherto been found in nature in actual chemical combination with gold. It also occurs in a native state, and, combined with other metals, forming tellurides. The most important of these are given below, but tellurides of mercury, bismuth, lead, and nickel also exist.

Tellurium has a bright tin-white color and a metallic lustre. It is brittle and very fusible, volatilizing almost entirely and tinging the blow-pipe flame green. White coating on charcoal. Soluble in nitric acid. Rare.

Nagyagite. Lead gray, very fusible, gives a blue color to the flame. Rare.

Hessite. Lead-gray, malleable, rare.

Petzite. Streak, iron black. Sometimes tarnished.

Sylvanite or *graphic tellurium.* Streak, steel-gray to silver-white. Color, steel-gray. Sectile. Gives the flame a greenish-blue color.

Calaverite. Streak, yellowish-gray. Massive, bronze-yellow, brittle, bluish-green flame.

The most common of these minerals, petzite and sylvanite, are of fairly common occurrence in Colorado, more especially at Cripple Creek.

(118)

Tellurides constitute exceedingly valuable ores when they are sufficiently rich to allow of hand picking and sale to smelters, and even the poorer ores can be treated by roasting and either chlorination or cyanidation. In many cases attempts to concentrate have been uusatisfactory, as the mineral frequently slimes a great deal; but concentration is said to have been successfully applied in Boulder County, Colorado, and the possibility depends to a great extent upon the nature of the ore. Specimens are found in many localities, but it is in comparatively few places that workable deposits exist.

PLATINUM occurs native and in flattened or angular grains or nuggets which are malleable. Its color and streak are steel-gray. Lustre metallic bright. Isometric, but is seldom found in crystals. Hardness 4 to 4.5. Specific gravity 16 to 19. As heavy as gold, and, therefore, easily distinguished and separated from lighter materials. Before the blow-pipe it is infusible : not affected by borax, except when containing some metal, as iron or copper, which gives the reaction. Soluble only in heated nitro-muriatic acid.

Platinum is occasionally found in the gold-bearing gravels of California and Oregon, but the annual production is small. There are no means of knowing whether it is present in sufficient abundance for separate mining. The prospectors, as a rule, do not know the value of the black sand, nor are they always able to distinguish it from less valuable ores; and it is, therefore, not unlikely that deposits may yet be found.

The supply of platinum comes chiefly from Russia, where it occurs in gravels, probably originally auriferous, on the Siberian side of the Ural. Since serpentine is usually near at hand, and the placers increase in richness as the rock is approached, and since the metal has been found in this rock, it seems probable that this is the source. This mode of occurrence of platinum and the association with serpentiferous rocks prevails also in other platinum-producing regions. Platinum is always alloyed with the other metals of the platinum group, iridium, osmium, palladium, etc., and with iron, the amount of platinum varying from 50 to 80 per cent. In Russia, as well as in other platinum-producing regions, chrome iron and iridosmium are associated with the metal. The United States now consumes more platinum than any other country, incandescent electric lamps and other electric apparatus calling for a great supply. Although only a very minute quantity is required in each case, so many lamps are called for that the demand is very great, and the price has risen much higher than formerly. It may be interesting to note that the name platinum is derived from *plata*, the Spanish word for silver, since it was regarded in South America at the time of its discovery (1735) as an impure ore of that metal.

Platinum, like gold, does not readily combine with other metals, and in nature the only compound known is an arsenide called *Sperrylite*, which is found in very small quantities in the Sudbury section of Ontario, Canada. Its color is tin-white; lustre bright; hardness about 7; specific gravity 10.6.

Platinum may be distinguished by its great weight, by its gray color, its sectile nature, and by the fact that it will not dissolve in any simple acid, and with difficulty in nitro-muriatic acid (aqua-regia). It may be distinguished from lead by its action under the blowpipe flame, since lead melts immediately, leaving a yellowish coating, while platinum refuses to melt under the hottest flame, and leaves no coating whatever. When it exists in the alluvial soil it may be "panned out" just as gold or other heavy metals, and even more easily because of its greater gravity.

It may be found in some metal-bearing veins in crystalline metamorphic and syenite rock, from which it has been washed down just as in the case of gold. In the latter condition it has been found more extensively than in any other.

Its chemical test is as follows: Dissolve the grains of the ore in nitro-muriatic acid (4 parts muriatic acid to 1 part nitric), preferably with gentle heat, add proto-chloride of tin (solution) also called stannous chloride ($SnCl_2$); if platinum is present a dark brownish-red color will be produced, but no precipitate.

The metal may be obtained separate from its gold, and in the presence of many other metals, by evaporating the above solution of the ore in a porcelain dish to dryness, at a gentle heat with ammonium chloride (sal ammoniac or muriate of ammonia), and the residue treated with dilute alcohol (one-fourth part water). The gold will remain in solution and the platinum be precipitated, the precipitate to be ignited, when the platinum will be pure. The gold, if present, may be precipitated by adding a solution of ferrous sulphate, after evaporating off the alcohol. Ferrous sulphate is proto-sulphate of iron (copperas in crystals).

Stannous chloride may readily be purchased at any chemist's warehouse, but as it is easily prepared we give the best method as follows : File a piece of tin into powder and heat very hot (nearly to boiling) with strong hydrochloric acid in a porcelain dish or beaker-glass, always keeping tin in the glass or dish, by adding tin if necessary. When no hydrogen gas is evolved (*i. e.*, no bubbles arise), dilute with four times its bulk of pure water, slightly acidulated with hydrochloric (muriatic) acid, and filter. Keep the filtrate in a well-stoppered bottle in which some tin has been placed. If you have pure tin-foil, that form of tin may be used, for without the presence of metallic tin the stannous chloride ($SnCl_2$) is in danger of changing into stannic chloride ($SnCl_4$) with precipitation of a white substance (oxychloride of tin), which renders the reagent unfit for use.

IRIDIUM, a steel-white, extremely hard metal, next in specific gravity to osmium, is supplied partly from its alloy with native platinum, and partly from the iridosmium which occurs in the platiniferous gravels. It is used for pen-points and in jewelry, and recently in metal-plating.

OSMIUM is the heaviest known metal. It comes from the same sources as iridium, and in the form of iridosmium is used for pointing tools and pens.

PALLADIUM is a brilliant, silver-white metal. It also occurs with platinum, but on account of its high price is but little used.

SILVER. This metal occurs native in various shapes, as in small grains in the rock, as branching and leaf-like, and also in small octahedral crystals and in other forms. Hardness, 2.3 to 3; specific gravity, 10.1 to 11.1, according to its purity. It is never found absolutely pure, but contains some gold and frequently a little copper.

It is always sectile and malleable, and in this respect very easily distinguished from a substance frequently mistaken for native silver, namely, *mis-pickel*, which is an *arsenide of iron*, having very much the appearance of silver, but always brittle.

BEFORE THE BLOW-PIPE, on charcoal, native silver is distinguished from tin, zinc, antimony, or bismuth, by the fact that it melts and leaves no whiteness or any other appearance of oxide upon the coal around the globule.

Tin will leave a white film, and lead a yellow; zinc a yellow which whitens on cooling. But silver leaves no film or cloud of any kind upon the coal.

The CHEMICAL TEST of silver is as follows: Dissolve the metal in nitric acid in a test-tube, preferably with the heat of an alcohol flame, but not to the boiling point. Add an equal amount of pure water (clear rain water will answer), then drop in several drops of a solution of common table salt or muriatic acid. If a cloudy white precipitate occurs which settles and blackens after exposure of a few seconds to sunlight or a few minutes to daylight, the substance is silver.

It should be remembered at this point that this test is for silver alone, since lead and mercury are also precipitated as a white cloud by the same solution, but neither blackens by exposure to the light. This distinguishes silver. If, however, further proof is needed, drop into the test tube strong ammonia water; the precipitate is dissolved if it is that of silver; it is not if it be of lead, and it is blackened by the ammonia if it is mercury.

If there is much copper in the silver it may be detected by dipping a clean strip of polished iron or steel into the solution, for the metallic copper will immediately appear upon the surface of the iron.

It must not always be supposed that native silver is metallic or white in appearance, for it is readily tarnished by sulphur, and the proximity of sulphur in other minerals or in water may greatly discolor the native silver.

Comparatively speaking, very little of the silver of the mines is derived from native silver. Most of the silver of commerce is obtained from some of the minerals named below, which are combinations of silver with other metals, and with sulphur or chlorine, as sulphides of silver, etc., in which condition they bear no resemblance to native silver.

But in all silver minerals of any commercial value, the already mentioned tests are usually sufficient to detect the existence of silver.

Other forms in which silver is found are—

SILVER SULPHIDES are very largely associated with lead sulphides or galena, and sometimes called, when pure, *silver glance* or *argentite*. This is found in masses, but when crystallized it occurs in cubes or octahedral forms. When freshly broken it has a metallic lustre, otherwise it is of a dull gray or leaden appearance. It is sectile, and its " streak " or the color of its powder is the same as that of the mineral itself, and rather shining. Chemical composition : silver 87 ; sulphur 13. Hardness 2 to 2.5. Specific gravity 7.1 to 7.4.

The ore is soluble in nitric acid, and on adding common salt to the solution a white curd is thrown down which blackens on exposure to sunlight. It is very fusible, giving off an odor of sulphur when heated. Before the blow-pipe on charcoal, with or without carbonate of soda, it yields a white globule of metallic silver which can be flattened under a hammer.

The ore occurs in veins in granite, porphyry, and slate, with arsenic, silver and lead ores.

HORN SILVER (*Cerargyrite* is the mineralogical name). The mineral known under these names is a chloride of silver occurring in massive form and sometimes in crystals. It has a resinous lustre and yields a shining streak. It is translucent on the extreme edges, and has a waxy appearance. It cuts like horn or wax, and on an outcrop looks like dirty cement. It contains 75.3 per cent. silver, and 24.7 per cent. chlorine when unmixed or nearly pure, and then has a pearly-gray or greenish-gray appearance.

A polished piece of iron may be slightly coated with silver if a piece of horn silver is moistened and rubbed upon the iron.

Horn silver is very easily fusible, it melting in the flame of a candle. Heated with carbonate of soda on charcoal, it yields a globule of metallic silver.

This mineral, in various degrees of impurity, forms a very large part of the silver-bearing ores of some mines in South America, as well as in the Western States and Territories of the United States. It is a valuable ore.

BRITTLE SILVER ORE (*Stephanite* is the mineralogical name) is a *silver sulphide with antimony,* and is found in masses and sometimes in rhombic prism crystals. It is easily distinguished from silver sulphide (or glance) by the fact that it is brittle, while the glance, if fairly pure, may be cut with a knife in chips without breaking.

This ore is black or iron gray, has a hardness of 2 to 2.5 and a specific gravity of 6.2 to 6.3, and when pure, contains 71 per cent. of silver, the rest being antimony with some other admixtures, usually iron

or copper. It is an abundant silver ore in the
Comstock Lode, Nevada (Figs. 41, 42), in the Reese
River and Humboldt and other regions, and at the
silver mines in Idaho.

On charcoal, under the blow-pipe, it decrepitates
and coats the coal with a film of antimony (anti-
monous acid), which, after considerable blowing,
turns red, and a globule of silver is obtained.

RED SILVER ORE, or RUBY SILVER, is an ore
which contains arsenic and antimony, or more usually arsenic or antimony. That containing only
antimony is dark red and is known mineralogically as PYRARGYRITE; it contains 59.8 per cent.
silver, 17.7 per cent. sulphur, and 22.5 per cent. of
antimony. It occurs generally in crystals. When
the silver sulphide is associated with arsenic only,
the color is light red and the name PROUSTITE is
applied to it. It contains 65.5 per cent. of silver.
It may contain both arsenic and antimony, and
have a grayish appearance. In Idaho, it has been
found in masses of several hundred pounds weight,
at Poorman Lode (Dana). In Mexico it is worked
extensively as an ore of silver.

BROMIC SILVER or BROMYRITE. This is a common ore containing bromine 42.6 per cent. and
silver 57.4 per cent.

There are other minerals in which silver occurs,
but they are only exceptions or rare, and if one is
acquainted with those mentioned above, he will
very likely detect the rarer silver minerals, which
are not ores in the usual sense, but they may lead
when discovered to valuable results.

Valuing silver ores. A simple, but rough, method is sometimes adopted of testing the value of ores from day to day when chlorides are the minerals chiefly worked, by powdering the ore in the mine, mixing it with a solution of hyposulphite of lime which dissolves the chloride, and then adding sodium sulphide, which forms a dark-colored precipitate if much silver is present. It is evidently impossible to estimate in this way the contents of silver, but it affords a very good test whether the ore is of value or not.

GEOLOGY OF SILVER ORES. The most valuable ores occur in the earlier or more ancient rocks, such as the granitic or gneissoid rocks, clay slates, mica schists, older limestones, and in the metamorphic rocks. The remarkable geologic conditions under which silver ores and veins occur may be understood more readily by the following diagrams than by any descriptions without them. (Figs. 41 and 42.)

In the diagrams the rocks are seen tilted up from the horizontal position to one nearly vertical, but evidently after this uplifting the trachytic dykes were shot through the masses of conglomerate. The lodes bearing silver are represented by continuous double lines, and the dykes by dotted vertical lines. The entire distance represented from Sutro to the west end of the diagram is about $5\frac{1}{2}$ miles, on a course east and west, being the same as that of the Sutro tunnel upon this branch, which joins or intersects to the north and south branch of the tunnel at the Comstock lode.

In order that the superficial nature of the country may be understood, we have given the north and south section of the same region, showing some of the mines by vertical black lines and by shaded spaces where the mines have been worked more or less extensively. (Fig. 42.)

The north and south section exhibits the hilly surface, and fully illustrates the work of the prospector who would become acquainted with the mineral deposits of a similar region.

It will be seen in the east and west section that all the lodes out-crop. (Fig. 41.) The non-metallic substances of these lodes are quartz, fluorspar, with, perhaps, some chlorides or sulphides; the latter may be metallic, and there may occur some traces of gold and silver, perhaps also of antimony, lead, etc. The wisest course, therefore, is for the prospector, after having settled in which direction the strike or course of the strata runs, to make an examination directly across the strata, the chief object being to learn the nature of the rocks of the region, and, at the same time, to detect the outcropping of any lodes or dykes.

His object is to become acquainted with the strata by means of the loose material, the fragments, or small outcropping rocks, where he cannot penetrate beneath the soil.

It may become necessary to traverse a great distance before any certain information may be gained, and where the hill surfaces are covered with soil, the ravines will frequently disclose the nature of the rock.

9

FIG. 41.

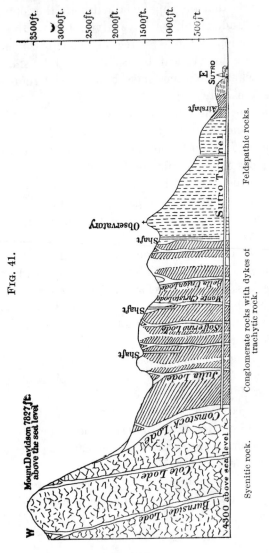

Syenitic rock.   Conglomerate rocks with dykes of   Feldspathic rocks.
                        trachytic rock.

SECTION ACROSS THE COMSTOCK LODE AND SURROUNDING STRATA, EAST AND WEST.

It will be noticed that the Comstock Lode begins immediately adjoining the syenite rock, and at the outcrop extends six or eight times the actual thickness of the lode below. It is also apparent that the lodes generally, at least in this region, bifurcate near the surface, even in the syenite, and when an outcrop has been discovered, the probability is that not far off another outcrop of the same lode may be found (Fig. 41).

The Comstock Lode has been traced for four or five miles north and south, but the values of the deposits are not uniform. The great bodies of ore may be seen in the north and south section where the excavations are largest, as around the Savage, and from the Exchequer to the Crown Point properties. But this whole region is filled with dykes and lodes for miles beyond the Comstock Lode, which lies on the eastern slope of a range of hills running somewhat parallel, but about fifteen miles east of the great Sierra Nevada range, south of the Pacific Railroad, and between the lakes Bigler and Carson in the western part of the State.

In the east of Nevada, at the Eureka Mines, the ores are found in a bed of limestone overlying the granites, quartzose slates, and metamorphic rocks of great thickness. The limestone containing the ore is about 300 feet thick. But while the immediate geology varies from that of the Comstock, the general facts are the same, namely, that the silver-bearing lodes are in or very near the granites or earliest rocks. In this case the overlying rocks,

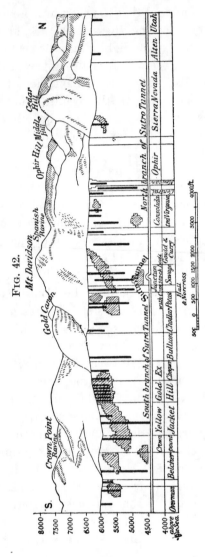

NORTH AND SOUTH SECTION OF THE COMSTOCK LODE, SHOWING THE MINES AND THE SURFACE.

though limestone, are dolomitic, containing from 36 to 46 per cent. of carbonate of magnesia, and the mineralized belt of limestone, or that containing the ores, is very much broken, and in some places apparently crushed, as if it had been subjected to a grinding process, and then partly rejoined by the cementing power of calcareous matter deposited from solution in percolating water.

A peculiarity in this last described limestone is found in the large caverns which occur along the course of mineral deposit. On the floors of these caverns are found beds of ore which seem to have dropped from their position in the limestone, as that has been dissolved out and carried off where the fissures easily permitted the percolating waters to pass rapidly away.

The geology of this region appears to be in the order of granites, quartzose slates, and metamorphic rocks of great thickness, limestones containing segregations of ore, calcareous shales, and these surmounted by limestones also of great thickness. The special region to which this geological series refers is in the Ruby Hill mines.

The Emma Mine, with many others, is situated still further east, in the Wahsatch range of moun- tains, which runs north and south about twenty miles east of the Great Salt Lake. This mine is about the same distance southeast of the Great Salt Lake. The adjacent rocks of this mine are granite. in massive beds dipping from 50° to 70° eastward, This is overlaid by quartzites of a reddish color,

then occurs a series of slates, upon which are thick
beds of white limestone, and these pass rapidly into
the carboniferous dolomitic limestone. It is in
this last limestone that the ore deposits of the
Emma and adjacent mines are worked.

It is a fact, however, that the ores are mainly
composed of silica and lead, of which there is over
70 per cent. The amount of silver is about 0.40 to
0.50 of 1 per cent. according to some analyses. A
sample amount of 82 tons, gross, yielded 156 ounces
of silver.

These three mining districts present the general
geologic conditions in which the silver ores are
found in these and other States and Territories, and
the prospector should expect to find surface indications accordingly, but modified more or less by exposure to weather.

Although, from the preceding illustrations, silver
is shown to be found both in the very early groups
of rocks and in the carboniferous limestone, the
latter is the exception, as it appears to be found
there only when that limestone has occurred with
little or no separating horizons from the earliest
rocks.

CHAPTER VIII.

COPPER. It occurs both native and in a compound state. Native copper is found in various forms, and even in octahedral crystals. Its color is copper red; it is always sectile and malleable; hardness 2.5 to 3, specific gravity 8.5 to 8.9, according to purity. Frequently associated with native silver. It is tested by the *blow-pipe ;* giving in small quantities blue tinge to almost black in the borax bead, according to quantity used, and the kind of flame, whether inner or *R*, or outer or *O*, the latter giving blue color, the former giving the copper color or metallic opaque brown.

Chemically, it dissolves readily in nitric acid, and, if ammonia be added, the solution becomes green, or greenish-blue if ammonia be in excess.

In the absence of any chemicals or a blow-pipe, the mineral, when containing native copper, or when only a compound containing copper, may be tested by heating it either in the mass, or, better, in powder, and when hot, dropping it into some salty grease and then putting it in a flame or upon burning charcoal, when the characteristic green color will appear in the flame with great distinctness.

(135)

Moreover, if the mineral contains copper in considerable quantity and it is dissolved in nitric acid, the copper will be deposited immediately upon a strip of polished iron or upon the end of a knife blade, if either be dipped into the solution.

The natural combinations of copper are almost endless. Not less than a hundred mineral species may be regarded as copper ores from the practical miner's point of view, *i. e.*, possessing economic value, and there are probably as many more which are not yet utilized. As might be expected the range of chemical associations is equally wide, embracing sulphides, antimonides, arsenides, oxides, chlorides, bromides, iodides, carbonates, sulphates, phosphates, silicates, arseniates, simple and compound, hydrated and anhydrous, in almost every degree of variety.

Below several of the more important ores of copper are mentioned, and also some copper minerals which, to the prospector, will be suggestive that the more important ores are not far off.

RED COPPER ORE OR RUBY COPPER (*Cuprite* is the mineralogical name) : Occurs massive, granular, and earthy ; brittle ; if in crystals, octahedral and twelve-sided ; nearly opaque ; deep red or ruby colored, sometimes weathered to an iron-gray on the surface ; hardness 3.5 to 4 ; specific gravity 8. Composed of copper 88.78 per cent., the remainder oxygen, when pure.

Before the blow-pipe, on charcoal, it yields a globule of metallic copper ; with borax bead gives the indications of copper. It forms a blue solution in nitric acid. These tests distinguish it from red oxide of iron. It occurs in granite and slate with copper ores and galena, and forms a valuable source of the metal.

COPPER GLANCE *or* VITREOUS COPPER (*Chalcocite* is the mineralogical name): Massive; slightly sectile; color blackish-gray, tarnishing to blue or green. Hardness 2.5–3; specific gravity 5.5–5.8. Composed of copper 77.2; sulphur 20.6, and sometimes, a little iron. It is fusible in a candle flame.

Before the blow-pipe it gives off an odor of sulphur. When heated on charcoal, a malleable globule of metallic copper remains, tarnished black, but rendered evident on flattening under a hammer. With borax bead it gives the indications of copper. Dissolves in nitric acid, forming a blue solution. These tests distinguish it from sulphide of silver. Occurs with other copper-ores.

GRAY COPPER (*Tetrahedrite* is the mineralogical name) : Brittle; steel-gray or iron-black, sometimes brownish ; hardness 3–4; specific gravity 4.75–5.1. Composed of copper 38.6, sulphur 26.3, and frequently antimony and arsenic, zinc, iron, silver, etc. It frequently contains silver, and sometimes as much as 25 to 30 per cent. Before the blow-pipe it gives a bead of copper or of copper and silver. It occurs with copper pyrites, galena and blende. This ore is wrought for copper and occasionally for silver.

COPPER PYRITES (*Chalcopyrite* is the mineralogical name): Massive. Color is a brass yellow, sometimes tarnished and iridescent. Hardness 3.5 to 4; specific gravity 4.15. Composed of copper 34.6 ; sulphur 34.9 ; iron 30.5. Before the blow-pipe it fuses to a magnetic globule on charcoal, and with borax metallic copper is the result. It is sometimes mistaken for gold, or iron, or tin pyrites. But it is brittle, while gold is not; it will not strike fire as does iron pyrites; and it may be distinguished from tin pyrites by the film that the latter leaves on the charcoal, while copper pyrites leaves no residue under the blow-pipe. It occurs in granite and slate in lodes or veins, and is a valuable ore of copper.

What is called *peacock ore* is only pyrites coated with oxide and exhibiting iridescent colors. By leaving a piece of clean yellow copper pyrites in water for some time it will become coated in this way.

SILICATE OF COPPER (*Chrysocolla* is the mineralogical name) is a bright-green or bluish-green mineral, scarcely worthy of being called an ore, although it contains from 35 to 40 per cent. copper and a large amount of silica. It is a secondary deposit. Its hardness is 2 to 4, and specific gravity 2 to 2.3. Its only significance to the prospector is that it may be associated with true ores. Its powder (streak) is white, while the mineral itself is green ; this being due to the quartz or silex in it. It does not entirely dissolve in nitric acid. Before the blow-pipe with soda, it gives a bead of copper.

BLACK OXIDE of copper is usually found on the

surface, and is generally due to the decomposition of some sulphide or other copper ore. It occurs in masses of a dark, earthy appearance, and sometimes in minute shining particles, and soils the fingers when handled.

MALACHITE, *green carbonate of copper*, has a fibrous structure nearly opaque, and of an emerald-green color, and contains about 57 per cent. of copper. Hardness 3.5 to 4 ; specific gravity 3.6 to 4.

Before the blow-pipe it becomes blackish. With borax it yields the usual blue-green bead, and on charcoal is reduced to metallic copper. It completely dissolves in nitric acid, and thus it may be distinguished from silicate of copper, which has nearly the same color and will not dissolve.

BLUE CARBONATE of copper (*Azurite* is the mineralogical name) is only used for ornamental purposes. It is of a deep blue color, sometimes transparent, brittle, and gives a bluish streak. It has a hardness of 3.5 to 4.5 and a specific gravity of 3.7 to 4. Can be scratched with a knife. It blackens when heated. On charcoal it is reduced to a globule of pure copper. With the borax bead it gives the indications of copper. It is soluble in nitric acid with effervescence, forming a blue solution.

VARIEGATED COPPER PYRITES (*Bornite* is the mineralogical name, but is also called *Erubiscite*) : Usually massive, of a copper-red to a pinchbeck-brown color, and a blackish to lead-gray streak. Hardness 2.5 to 3, specific gravity 5.5 to 5.8. It contains 79.8 per cent. copper and 20.2 per cent. of sulphur. Before the blow-pipe it gives a bead of copper.

But the minerals above mentioned are by no means the most important as regards the commercial supplies of the metal; in fact in that light they may almost be disregarded so far as affording any considerable proportion of the total yearly output, though, of course, deposits of these ores are profitable. The bulk of the world's consumption of copper is furnished by ores of the lowest grade, ranging from little more than $\frac{1}{2}$ to perhaps 5 per cent., though rarely more than 3 to $3\frac{1}{2}$ per cent. Thus the ores of Devon and Cornwall are worked for $1\frac{1}{2}$ to 2 per cent. copper; those of Cheshire, for less than $1\frac{1}{2}$ per cent.; those of Mansfeld, Germany, for little over $2\frac{1}{2}$ per cent.; those of Rio Tinto, Spain, for $2\frac{1}{4}$ to $3\frac{1}{2}$ per cent.; those of Maidenpec, Servia, for 2 to 3 per cent.; and, overwhelmingly the most abundant producers, those of the Lake Superior region for as little as 0.65 per cent.

Formerly the world's supply of copper was drawn from the rich ores, containing up to 40 per cent. of metal as mined, and further explorations may again reveal in the future similar deposits to replace those now exhausted; but at present and in the immediate future reliance must be placed on the enormous low grade ore bodies now being worked, especially in North America.

The GEOLOGY OF COPPER is more varied than that of many other metals, as it occurs in rocks of almost

every age. In Cornwall the slates are more productive than the granites, while in our mines in the
Eastern States the new red sandstone, the carboniferous limestone, and silurian rocks furnish copper.
Also found in the metamorphic limestone, near
slate (Fig. 43). In the Lake Superior region, where
large deposits of native copper are found, the rocks
are sandstones and shales underlying green-stone or
a kind of trap, and in some places seem to be igne-

FIG. 43.

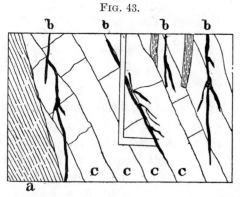

SECTION OF THE COPPER BED AT THE DOLLY HIDE MINE, MARYLAND. *a*,
Slate, *b, b, b, b*, Ore beds or segregations of ore. *c, c, c, c*, Crystalline limestone (metamorphic).

ous (Figs. 44, 45). Ruby copper ore occurs in Arizona between quartzose and hornblendic rocks and
limestone. It occurs in both, lodes and deposits,
and the best way for the prospector to prepare for
actual discovery is to make himself well acquainted
with the copper compounds, whether ores or minerals. They may indicate true ores, although they
contain little copper.

To become ready in the detection of copper as an ore the following facts should be kept in mind, as furnishing suggestions for skillful practice. (Figs. 43, 44, and 45.)

It is well to remember, especially when exploring a new country, that copper is frequently associated with rocks of a dark color, which are very often

FIG. 44.

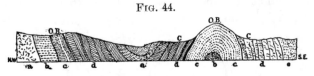

Section of strata in Lake Superior copper region : *a*, Granite. *b*, Gneis soid. *c*, Greenstone, hornblende, conglomerates with interstratified slates *d*, Slaty rocks and traps, etc. *e*, Potsdam sandstone. *C*, *C*, Places of copper deposits. *O*, *B*, Iron ore beds. Section from *N. W.* to *S. E.*

green ; but it must not be supposed that the color is imparted by copper, for it is generally due either to some other metal, such as iron, or to the presence of a green non-metallic mineral, such as chlorite. Serpentines and hornblendic rocks are often associ-

FIG. 45.

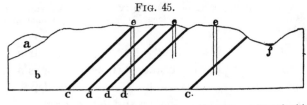

Copper. Section of the Eagle vein, Lake Superior. *a*, Poryphyritic rocks. *b*, Greenstone. *c*, *c*, Conglomerate. *d*, *d*, *d*, Amygdaloid bearing copper. *e*, *e*, *e*, Shafts. *f*, Montreal River.

ated with copper ores, but green serpentines owe their color to iron, nickel or chromium, and if copper is found disseminated through some of them, it is the exception and not the rule, unless in the immediate vicinity of ore deposits. On the contrary, iron and chromium are found in all serpentines, and nickel is of frequent occurrence.

All copper ores weigh more than quartz or limestone, and the comparative weights should be so well known by practice that there should be no hesitation in judging that the mineral you hold is more than 2.6 in specific gravity, 2.6 being that of either quartz or limestone.

Next examine the mineral with your pocket lens for any evidence of copper, such as green or bluish spots, or brassy points or particles; if found, chip one off and use the blow-pipe with borax bead or with soda or borax on charcoal. If the characteristic color appears, it is copper. Now proceed with other parts of the specimen. If a sulphury smell is plain, it is probably a sulphide. Place a small chip upon a depression in the charcoal, cover with soda or borax, turn the inner flame upon it and reduce to a metallic globule; if it shows the color of copper and is malleable, it is copper; if it blackens, apply your magnetized knife-blade, and if it is attracted the mineral contains iron, and it may contain both iron and copper.

The next work is to examine the region to gather any other specimens and evidences of true ores, before attempting to know more of any particular specimen. If the surface specimens are numerous it may be well to gather some six or eight and proceed to an examination as to the available copper. This is now the work of the chemist, and should be submitted to him   But as the skillful prospector frequently wishes to be his own chemist, where work for the desired object is not difficult nor very complicated, we give the following simple process of arriving at the per cent. of copper in an ore without regard to other elements contained therein :

To OBTAIN THE PER CENT. OF COPPER IN AN ORE. The only chemicals needed are nitric acid, ammonia, and sodium sulphide—the colorless crystallized hydrosulphide of soda of commerce is good enough. All the apparatus needed is a glass flask or tall beaker-glass and a marked tall glass called a burette. This glass may be obtained at any chemical warehouse. The burette is marked in cubic inches or cubic centimetres, from 25 to 100. Dissolve some sodium sulphide in clear rain-water—about a half ounce to a pint. Keep the solution in a glass-stoppered bottle. Obtain some pure copper (ordinary good copper wire will answer), weigh the piece accurately and dissolve in nitric acid, add some water (twice the amount of acid used, or a little more), then add ammonia until, when stirred with a long piece of glass or glass rod, the solution smells strongly of ammonia. The ammonia must be in excess. Now fill the burette with sodium sulphide to the 100 mark, and from the burette pour into the copper solution until the blue color of copper entirely disappears ; note on the burette by its marks the exact amount of sodium sulphide used. That amount represents the weight of the amount of copper used.

Now for the ore : Pulverize some of the averaged ore, weigh it, and treat it as you did the copper with nitric acid and ammonia, and proceed with the sodium sulphide. When the ore solution has become entirely colorless, note what amount of sodium sulphide solution you have used, and you may then calculate the exact amount of copper in the ore by simple proportion. The presence of tin, zinc, lead, iron, cadmium, antimony, arsenic, or bismuth in the ore does not interfere with the operation. But silver does. Therefore, a small amount of the ore must be dissolved in nitric acid (free from all muriatic acid or chlorine, as this would precipitate the silver before you would notice it), and tested by dropping into the solution a drop or two of hydrochloric acid or solution of common table salt (sodium chloride). If any silver exists in the ore a milky cloudiness will appear, of a density greater or less in accordance with the amount of silver present. If no silver appears, then you may proceed as already directed. If silver does appear, then the solution containing the weighed ore must first be treated with the salt solution or diluted hydrochloric acid, until all cloudiness or white precipitate entirely ceases. The solution of ore now contains no silver, and you may proceed as directed.

This process is sufficiently accurate for all assays provided the following precautions are observed :—

1. Heat the copper solution, after adding the am-

10

monia, to boiling point or little below while adding the sodium sulphide. 2. Add a little ammonia to the ammoniacal solution to keep it from losing ammonia by evaporation. 3. When the blue ammoniacal solution begins to lose its color, drop the sodium sulphide in cautiously, so as not to exceed the amount necessary to exactly precipitate the copper and no more.

Note the precipitates : The sodium sulphide first produces its black precipitate of copper sulphide, but before that takes place the ammonia will produce another precipitate, provided the copper contains any lead or tin. If the copper contains zinc, that will be precipitated immediately following the black copper sulphide, but will be white. If it contains any cadmium, that will be precipitated at the very moment the decoloration takes place, if the adding of the sodium sulphide is continued. Cadmium is known by a beautiful clear yellow precipitate. With care and skill each may be noticed.

In simply determining the amount of copper, however, no regard need be had to any of these precipitates, only pay attention to the point of decoloration.

The sodium sulphide may need proving to see if it has lost any of its strength if kept for a long time, and this may be done by a trial with a new solution holding a known amount of copper. Or, exactly the same weight of crystals of sodium sulphide to the same amount of pure water may be used as before, and the old solution thrown away. Or, by re-testing the sodium sulphide the same solution may be used for a long time, and if it has become weakened, make allowance for the additional sodium sulphide required. It should be kept in a cool place, out of the sun and light also.

CHAPTER IX.

LEAD. It very rarely occurs native; it then has a hardness of 1.5 and specific gravity 11.3 to 11.4. But the most usual ore of lead is the sulphide called GALENA. When chemically pure it contains 86.55 lead and 13.45 sulphur. Its specific gravity is 7.2 to 7.6, according to admixtures. Streak, lead-gray. Color, metallic lead-gray. Easily recognized by the characteristic cubical cleavage which is very easily obtained, or granular structure when massive. Frequently associated with other metallic sulphides such as pyrite, chalcopyrite, arsenopyrite, blende, etc. It occurs in veins, the gangue of which is either quartz, calcite, barite or fluorspar, in granite and nearly all varieties of rock, but the larger deposits are usually found either in veins or in pockets, often of great size, in limestone strata.

Galena almost always contains silver, and hence all galenas should be tested for silver.

TEST FOR SILVER IN GALENA. Powder the galena and dissolve it in strong nitric acid (fuming acid is best, which has been described), then dip a piece of polished copper strip in the solution, and, if silver exists in any amount, there will be formed a film of silver on the copper. If the thin film be.

(148)

comes decidedly silvery, and in a short time, the ore should be laid aside for a more careful analysis.

The order of strata in the galena district of Wisconsin, Illinois and Iowa is shown in the annexed table.

	NIAGARA LIMESTONE.
CAMBRO-SILURIAN	⌠ Galena limestone which bears lead. ｜ Trenton limestone, fossils. ⟨ Sandstones, shales, and calcareous beds. ｜ Lower magnesian limestones. ⌡ Lower limit of lead.
	WHITE POTSDAM SANDSTONE.
CAMBRIAN	⌠ Upper— ｜ Fossiliferous slates. ⟨ Lower— ｜ Dolomitic limestones. ⌡ Dark sandstones.

Order of Strata in the Lead District of Wisconsin, Illinois, and Iowa.

The geology and form of lodes of the galena ores are seen in Fig. 46.

FIG. 46.

Lead Lode in Micaceous Slate in Mine near Middletown, Conn.

In several regions, but very extensively in Colorado, a rich carbonate of lead has been found (Fig. 47).

CARBONATE OF LEAD (*Cerussite*, mineralogical name.) If perfectly pure, its composition is, lead 83.6, carbonic acid 16.4. As a mineral its hard-

FIG. 47.

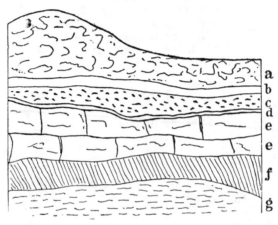

SECTION OF STRATA IN CALIFORNIA GULCH, COLORADO, SHOWING PORTION OF THE CARBONATE OF LEAD DEPOSITS. *a*, Porphyritic rock, 12 to 100 ft. thick. *b*. Thin bed of white clay. *c*. Carbonate of lead bed, 1 to 20 ft. thick. *d*, Oxide of iron; 1 to 6 ft. thick. *e, e*, Limestone. *f*, Clay slates. *g*, Quartzites and metamorphic rocks resting upon gneiss.

ness is 3 to 3.5, its specific gravity 6.4 to 6.5. Color (if freshly broken), white to gray, or even black, if it has been much weathered. If in good condition, it is translucent, or even transparent. Very brittle. If it contains copper it is usually tinged blue or green. It has a glassy or vitreous appearance, and is easily melted before the blow-pipe, and a lead bead or globule is readily obtained.

By using a little bone-ash plastered in a hollow in the charcoal and turning the *O F* upon the lead, after a little skillful blowing the lead is absorbed and drawn off and a bright silver globule remains, provided the lead contains silver. This is *blow-pipe cupelling*.

SULPHATE OF LEAD often accompanies the carbonate. It somewhat resembles the carbonate, although it is of slightly less hardness, 2.75 to 3, specific gravity 6.12 to 6.3. It may be distinguished from the carbonate by the fact that it *does not effervesce in an acid*, as the latter always will. Its mineralogical name is *anglesite*. It is composed of lead oxide 73.6 and sulphuric acid 26.4 in the pure specimens.

PHOSPHATE OF LEAD. Mineralogically, *pyromorphite*. Composition, when pure, 89.7 phosphate and 10.3 chromate of lead, with arsenate of lead (0 to 9), phosphate of lime (0.11), and fluoride of calcium. Hardness, 3.5 to 4; specific gravity, 6.5 to 7; color, green with modifications. It has a resinous lustre and is translucent; contains 78 per cent. lead. Heated on charcoal before the blow-pipe a globule is formed which takes on a crystalline appearance on cooling, leaving a yellow oxide of lead on the charcoal. With carbonate of soda in the reducing flame it yields a yellow globule. It is soluble in nitric acid.

CHROMATE OF LEAD is a yellow mineral containing protoxide of lead 68.15, chromic acid 31.85. Hardness 2.5 to 3; specific gravity 5.9 to 6.1. Color, various shades of bright hyacinth-red, streak (powder) orange-yellow. Lustre, vitreous. Translucent, and sectile. Mineralogical name is *crocoite*.

LEAD OCHRE, *massicot* mineralogically. This mineral occurs massive, as a compact earth of a sulphury-yellow or reddish-yellow appearance. It has a hardness of 2, a specific gravity of 8, and, when pure, 9.2. It is composed of oxygen 7.17, lead 92.83. Before the blow-pipe it fuses readily to a yellow glass, and on charcoal is easily reducible to metallic lead.

LEAD-ANTIMONY ORES. There are several compounds of lead with antimony, but they are never sufficiently plentiful to be considered as ores. One of these, *jamesonite*, contains small proportions of iron, copper, zinc and bismuth. It occurs in gray fibrous masses or small prisms, and is found in Cornwall associated with quartz and bournonite. Another of these compounds, *zinkenite*, resembles stibnite and bournonite, and occurs in an antimony mine in the Hartz.

THE GEOLOGY OF LEAD. Almost all the galenas and the carbonates contain silver, and some of the latter, as in Colorado, contain large quantities of silver. The geology of lead is very much the same as that of silver.

The ores are found in veins and lodes, and also in flats and beds, and in pockets (Fig. 48). The galenas occur in limestones, called the "galena limestones," a yellowish-gray, hard, compact, crystalline rock. The lowest horizon of lead ore in workable quantities lies above that of copper.

"The limestones and underlying schists are, for the most part, in a metamorphic condition, and there can be no difficulty, from the presence of porphyry above and the quartzites and gneiss below, in recognizing their position," * as in the Cambro-silurian system. It is supposed that the largest proportion of silver is contained in the ore derived from this geologic horizon.

FIG. 48.

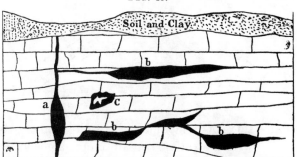

SECTION OF GALENA LIMESTONE showing how the lead occurs in lodes, *a*, flats, *b, b, b*, and pockets, *c*, from mere threads to several feet of thickness.

When water has had its course, however, the condition of a mine and of its veins and beds of ore may have been changed. Robert Hunt, as it regards British mines, says, that the circulation of water in the veins is affected by the inclination of

* B. C. Davies, F. G. S. A Treatise on Metalliferous Minerals, London, 1892, p. 259.

the strata in the direction of the vein.   The richest deposits are found in that portion of strata which is the most elevated, ·for instance, on the side of a powerful cross vein, thus:

The circulation of water is dependent upon an outlet at a lower level.

In the case of lead mines, it is stated that in consequence of the conditions connected with the descent of water, the richest deposits of lead are generally found at no great distance from the out-cropping of the containing rock.   Veins which run

FIG. 49.

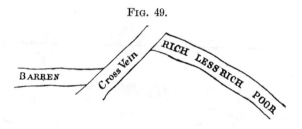

on the sides of a mountain in a direction nearly parallel with the valleys contain more extensive deposits of lead than those which cross the valleys at right angles.*

The prospector should keep this suggestion in mind.

The lead ores are found in the fissures where they seem to have been deposited by waters which have dissolved them out from neighboring beds (Fig. 50).

In the United States the chief sources of lead in

* British Mining, by Robert Hunt, London, 1884, p. 344.

late years have been argentiferous ores and considerable from zinc ores, but a notable exception is S. E. Missouri, where galena accompanied by nickeliferous pyrite is disseminated through magnesian limestone of Cambrian age. The mines are at Bonne Terre, Mine la Motte and Doe Run. The strata lie almost horizontal, and are known to carry lead through over 200 feet in thickness.

Tin. When a tin-bearing mineral is heated be-

FIG. 50.

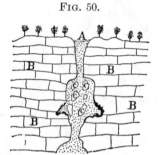

SECTION OF A LEAD DEPOSIT IN A FISSURE OF THE LIMESTONE. Williams & Co.'s Mine, Wisconsin. *B, B, B, B,* limestone. *A,* the fissure running down. *C, C, C, C,* masses of ore. Metamorphic.

fore the blow-pipe with carbonate of soda or charcoal, white metallic tin is yielded. By dissolving this in hydrochloric acid and adding metallic zinc, the tin will be deposited in a spongy form. In the blow-pipe assay tin leaves behind a white deposit which cannot be driven off in either flame. If it be moistened with nitrate of cobalt solution, the deposit becomes bluish green, and this test distinguishes it from other metals.

Assay of tin ore. If the ore is poor it should be concentrated, the vein-stuff being got rid of as much at possible. If mixed with iron or copper pyrites, it should be calcined or else treated with acids. One method is to mix the ore with one-fifth of its weight of anthracite coal or charcoal, and expose it in a crucible to a great heat for about twenty minutes. The contents are then poured out into an iron mould, and the slag carefully examined for buttons.

Another method is to mix 100 grains of the ore with six times its weight of cyanide of potassium, and expose the mixture to the heat of a good fire for twenty minutes. The contents are allowed to cool and afterwards broken to remove the slag.

The usual ore of tin is the oxide (binoxide) whose typical composition is tin 78.38, oxygen 21.62; hardness 6 to 7; specific gravity 6.8 to 7. It is, as a mineral, called *cassiterite*, and contains small quantities of iron, copper, manganese, tungsten, tantalic acid, arsenic, sometimes silica, and rarely lime. It occurs massive and in crystals, also in botryoidal and reniform shapes, concentric in structure and radiated fibrous, internally, and is then, in the last form, called "*wood tin*," from its woody appearance. *Toad's-eye* tin is the last described, but in very small shot-like grains, and *stream tin* is the same only in form of sand, found near or in streams.

Tin ore (binoxide) is nearly as hard as quartz, and will scratch glass, especially if freshly broken.

Pure crystals are rare. They are nearly transparent, but in the mass, as it occurs in the mines in Dakota and in many other places, the ore is a dark brown color, and sometimes almost black ; the fine powder or streak as made by a file, is light brown, however dark the mineral may be. The brown color or shade is due to oxide of iron in composition ; if perfectly free from all associated impurities it would be nearly white or colorless. The usual appearance in mass or pebbles, or finer, is that of a dirty or burned-brown color with varying depths of shade.

In the pebble form it is apt to wear quite smooth, due to its extreme hardness.

It was in this form that it was discovered in Banca, in 1710, and in the neighboring island, Billiton, and traced to its source in the mountains, where the central rock is granite, covered by quartzites, altered sand-stones, and slaty rocks. The altered sandstone just above the granite is the most productive rock, and it is traversed in all directions with tourmaline.* The same general associations largely exist in Wyoming and Dakota tin mines.

There is another mineral containing tin which may lead to the discovery of the true ore. It requires only a short description, which we give.

Tin pyrites (*sulphide of tin*) whose composition is, as a mineral, 29 to 30 sulphur, 25 to 31 tin, 29 to 30 copper, with iron and sometimes zinc. It has

* D. C. Davies, F. G. S., Metalliferous Minerals, London, 1892, p. 194.

been dug as an ore of copper and called "*bell-metal.*" Its hardness is 4; specific gravity 4.3 to 4.5 ; has a metallic lustre ; color, steel-gray to black, often yellowish from the presence of copper sulphide ; it is opaque and brittle.

With nitric acid. it affords a blue solution, and sulphur and tin oxide separate and may be tested on charcoal, where it fuses to a globule, which, in the oxidizing flame, gives off sulphur and coats the coal with white oxide of tin.

This ore or mineral, for it does not as yet deserve the name of tin ore, is of little use, but the prospector does well to make himself acquainted with it, as it is frequently associated with the binoxide or cassiterite, or black oxide, as the true ore is frequently called.

In the United States, cassiterite occurs in small stringers and veins on the borders of granite knobs or bosses, either in the granite itself or in the adjacent rocks, in such relations that it is doubtless the result of fumarole action consequent on the intrusion of the granite. It appears that the tin oxide has probably been formed from the fluoride. The Cajalco mine in California and the Harvey Peak mines, South Dakota, have been developed, but it is questionable whether they are worked at a profit. Undeveloped deposits are reported in Alabama, North Carolina and Virginia. At Broad Arrow, near Ashland, Alabama, tin-ore is disseminated in gneiss, the ore averaging about $1\frac{1}{2}$ per cent. black tin, but being very much mixed with titaniferous

iron. At King's Mountain, North Carolina, cassiterite occurs very irregularly in a "greisen" or
altered granite, and in limited alluvials derived
from the disintegration of the same. On Irish
Creek, Virginia, experimental parcels of vein-stone
taken from deposits in granite have shown $3\frac{1}{4}$ to $3\frac{1}{2}$
per cent. metallic tin, largely associated with arsen-
ical pyrites and ilmenite, which increase the difficulties of concentration and lower the value of the
product.

Tinstone stands nearly by itself in its mode of
occurrence and formation, as a type of a strongly
marked class of deposits. It is always associated
with granitic rocks, quartz-porphyries, or gneiss, all
of which are of analogous composition, being rich in
silica, which crystallizes as quartz, and being called
in consequence "acidic" rocks. Tin lodes are
nearly all of great antiquity and occur only in
those of the above-named rocks which are characterized by the presence of white mica. It is only
in two or three places in the world, notably Tuscany and Elba, that granites of this type have been
erupted during recent times, and they contain tin in
small quantity as well as some of the minerals
usually associated with it, such as tourmaline,
lithia, mica, and emerald.

Although this fact is of no immediate practical
value, it is important, because it shows that there
really are laws which govern the distribution of
minerals, although these are sometimes very ob-
scure ; but by constant observation it is certain that,

amongst discoveries of merely scientific interest, laws capable of practical application will occasionally be found.

Tinstone is always associated with quartz and rarely occurs in green rocks, unless their color be due to chlorite; nor in dark-colored rocks, except where stained red by the decomposition of ferruginous minerals; neither is it found in limestone.

Those granites which are characterized by abundance of white mica have, with good reason, been termed "tin granites," and a coarse-grained rock composed of granular quartz mixed with white mica and called "greisen" occurs in all the tin fields of the world.

The minerals most commonly associated with tin, namely topaz, mica, tourmaline, fluorspar, apatite and other rarer minerals containing fluorine, seem to show that it was originally contained in the granite as fluoride of tin, and that the associated minerals have been formed at its expense. It is an established fact in the genesis of minerals that fluorine is always accompanied by silicon and boron. It is therefore natural to find silicates containing boric acid, such as tourmaline and axinite, in association with tin. Other minerals which frequently accompany this metal are wolfram, molybdenite, mispickel, garnet, beryl, etc.

It is evident that a most important aid to the prospector is a study of the characteristics of the tinstone ores, and he may find it beneficial to become acquainted with the special minerals above mentioned as associated with the ores.

These minerals include, in some mines, *wolframite*, which gives trouble in the Cornwall and other tin mines, and the following description and tests may aid in detecting it :

Wolframite is in hardness 5–5.5, specific gravity 7.1–7.55, therefore, in these features it resembles the tin oxide ; though somewhat softer, yet the specific gravity is practically the same, although really heavier. So in color it frequently closely resembles tin oxide. But in the streak (or scratch powder), wolframite is a *dark reddish-brown* to black, while the tin oxide gives a white or grayish-brown powder ; wolframite is opaque, while the tin oxide is translucent and sometimes transparent *on the edges ;* when mixed with iron or manganese *rarely*, it looks almost opaque. Composition of wolframite : Tungstic acid about 75, the remainder protoxide of iron and manganese protoxide, more of the latter than of the former.

Wolframite is used in the preparation of some colors and enamels, and enters into the composition of some special kind of steel. Tungstate of soda which is used as a mordant and for fire-proofing fabrics, is also prepared from it.

11

CHAPTER X.

ZINC. This useful and well-known metal occurs in a variety of forms, its chief ores being:

ZINC CARBONATE. Mineralogical name *Smithsonite*; composition, zinc 51.44, oxygen 13.10, carbonic acid 35.46. But the composition in the mines varies because of the presence of protoxide of iron, manganese and magnesia. Color, when pure, nearly white, through various shades of yellow and gray to brown. Hardness 5; specific gravity 4–4.4. Streak, uncolored or white. Lustre, vitreous, pearly, subtransparent to translucent. Found in veins, but more usually in irregular deposits in limestone strata.

It is easily detected by the blow-pipe, as it gives a green color when heated after being moistened with half a drop of nitrate of cobalt solution. On charcoal, with soda, it coats the charcoal with a white film, which is yellow when hot and white on cooling, but if moistened with the cobalt solution and heated in the *O F* it turns green. With muriatic acid it effervesces and dissolves. In mass it is translucent and brittle.

ZINC SILICATE. Mineralogical name, *calamine;* composition, zinc oxide 67.5, silica 25, water 7.5.

(162)

Hardness 4.5–5, the latter when crystallized (Dana): specific gravity 3.16–3.9. Color and streak the same as in Smithsonite. Acts before the blow-pipe as does Smithsonite, but *does not effervesce with acids*, and gelatinizes; it is soluble in a strong solution of potash. In physical characters zinc silicate somewhat resembles zinc carbonate. An anhydrous variety of this ore is *Willemite*, which is found in New Jersey (Mine Hill and Sterling Hill). Zinc silicate is usually found in veins or in beds or in irregular pockets in stratified calcareous rocks, in association with zinc blende, zinc carbonate, iron, lead ores, etc.

RED OXIDE OF ZINC, mineralogical name is *zincite* (pron. zinkite), and its composition is zinc 80, oxygen 20, varied by the presence of 3 to 12 parts of peroxide of manganese, which gives the red color, for zinc oxide, pure, is white. The ore is peculiar to one region in New Jersey, Franklin, Sussex Co. Hardness 4–4.5; specific gravity 5.4–5.7; color, red and yellowish-red, streak the same; translucent, brittle.

SULPHIDE OF ZINC, mineralogical name *sphalerite* or *blende*, miners' name *black-jack*. Composition, zinc 66.8, sulphur, 33.2, but varied in the mines by iron, and sometimes cadmium. Color varies from yellow to brown and almost black, having a waxy look. Hardness 3.5 to 4; specific gravity 3.9 to 4.2; brittle, translucent. Zinc blende is the most abundant zinc ore. It occurs in rocks of all ages, in veins, in contact deposits or in irregular pockets in limestone, etc., and is frequently associated with the ores of lead, as well as those of copper, iron, silver, gold and tin; also, frequently associated with quartz, barite, fluorite, calcite, etc. It is easily recognized if treated with hot hydrochloric acid, as it gives a smell of rotten eggs (sulphuretted hydrogen), and the same results can be obtained without heating if a small quantity of pure iron filings is added to the acid. With soda on charcoal before the blow-pipe, zinc blende gives a sulphuret which, with water on a silver coin, tarnishes or blackens it.

The geology of zinc and of lead are so nearly alike that what has been said of the latter will apply to the former (Fig. 51).

FIG. 51.

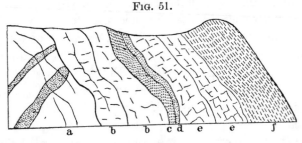

SECTION OF STRATA NEAR SPARTA, NEW JERSEY, ZINC MINES.

a, Slaty rock with feldspathic dykes. b, b, Limestone. c, Franklinite iron ore with zinc 20 to 30 ft. wide. d, Red oxide of zinc 3 to 9 ft. wide. e, e, Crystalline limestone. f, Feldspathic rock.

In New Jersey a section of strata near Sparta shows slaty rock with feldspathic dykes, then limestone adjoining the Franklinite iron ore with zinc 20 to 30 feet wide, then the red oxide of zinc 3 to 9 feet wide, then crystalline limestone, and next feldspathic rock (Fig. 51).

Enormous and extensive deposits of the sulphide are reported as occurring in Colorado, at George-town and Mount Lincoln, and in Montana, near Jefferson City.

The blow-pipe shows the same tests for zinc as have previously been mentioned. The fumes of sulphurous acid may be easily noticed when the mineral is placed in an open tube ·of glass (a test tube with a small hole in the bottom will be sufficient), and is strongly heated.

IRON. This metal is one of the most abundant and widely disseminated elements of the earth's crust, its distribution being materially aided by the fact of its forming two oxides of different chemical quantivalence. Native iron is not found in nature, but occurs with a small percentage of nickel in meteorites. It resembles ordinary iron, is malleable and attracted by a magnet. Specific gravity 7.0 to 7.8.

The chief ores of iron are magnetite, hematite (red and brown), and black band.

MAGNETITE is composed of iron 72.4 and oxygen 27.6. This ore is always easily attracted by the magnet, and sometimes is found capable of attracting iron, and then is called "*polaric*" or "loadstone."

Hardness 5.5 to 6.5; specific gravity 5 to 5.1. Color, nearly black; streak black. In powder or small grains it is always attractable by a magnetized knife-blade.

Nitric acid does not act upon it, but muriatic acid

dissolves it when in very fine powder, and under long-continued heat.

Iron exists in magnetite as protoxide and peroxide or FeO and Fe_2O_3, and upon this difference of oxides is based the action of important tests.

FRANKLINITE is an ore somewhat resembling magnetite in color, hardness, and specific gravity, but it contains manganese and zinc, and as an ore, is peculiar to Sussex Co., New Jersey. Its streak is dark brown, and its action on the magnet is feebler than in the case of magnetite. The iron is said to be of the composition of peroxide, or Fe_2O_3, but it is probably in part protoxide, and this is the cause of its feeble effect on the magnet.

It is easily affected under the blowpipe. Alone, it is infusible, but with borax in the $O F$ it colors the borax bead with the amethystine color of manganese, and in the $R F$ it shows the bottle-green of iron. On charcoal with soda it gives the bluish-green manganate, and also the coating of zinc, especially if the soda is mixed with borax. It is soluble in fine powder in muriatic acid.

SPECULAR ORE is the peroxide of iron without the protoxide. This oxide is also called the sesquioxide, or one and a half oxides, since iron combines with oxygen in the proportion of one to one and a half parts, or Fe_2O_3, and this is the highest proportion of oxygen the iron will combine with, and hence it is the peroxide, the peroxide and sesquioxide being the same in this case.

Specular ore is called RED HEMATITE from its

color, which in some masses is so intensely red as to appear nearly black, but it may always be distinguished from magnetite by its red streak, and the blacker the ore the more decided is the red of its powder or streak. It is never magnetic. We have always found that in cases where specular ore showed any magnetic attraction, it was due to the fact that the ore contained some protoxide of iron.

FIG. 52.

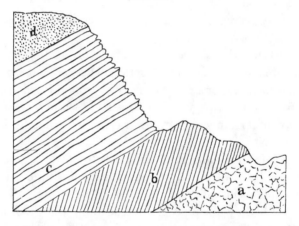

GEOLOGIC HORIZONS AROUND THE IRON ORES OF LAKE SUPERIOR.

a, Gneiss. *b,* Hornblende slates. *c,* The same with numerous thin beds of iron ore which frequently unite. *d,* Potsdam sandstone.

Hardness 5.5; specific gravity 4.5 to 5.3; composition, 70 per cent. iron, 30 per cent. oxygen. Color, reddish to almost black.

BROWN IRON ORE OR BROWN HEMATITE OR LIMONITE. This is the same composition as red hematite, except that it has less iron and contains

water in chemical combination, generally about 14
per cent. Color always brown. When heated red-
hot it loses its water and turns to a bright-red, unless
largely mixed with alumina and silex, when the
red color is shaded. It is not magnetic unless
heated with soda under the blow-pipe, when it becomes metallic, as all iron ores do.

The amount of metallic iron in a pure specimen
is 59 per cent., sometimes decreased by the presence
of alumina, silica, magnesia, and other impurities, so
that its average in many good mines is only about
35 to 36 per cent. iron.

SPATHIC IRON ORE OR SIDERITE is an iron carbonate, composed of iron protoxide 62 per cent. and
carbonic acid, or 48 per cent. pure iron. Hardness
3.5 to 4.5; gravity 3.7–3.9; streak white. Color
gray or cream color, unless weathered, when it is
brownish.

When in powder it effervesces with muriatic acid,
especially when hot. Translucent on edges, and
thin plates or splinters.

With the blow-pipe in a closed tube (test tube) it
decrepitates, becomes blackened, and gives off carbonic acid. Before the blow-pipe alone, held by
forceps, it blackens and fuses. In the test-tube with
muriatic acid it may be tested for carbonic acid, by
letting a lighted thread down into the tube, when
the flame is instantly extinguished. The solution
in the tube may be tested for iron by dropping a
drop of solution of ferricyanide of potassium into the
muriatic acid solution, when it becomes instantly a

deep blue. This is a test of protoxide of iron, spathic ore being iron in the condition of protoxide only.

BLACK BAND ORE is an argillaceous spathic ore of various dark colors, being largely combined with carbonaceous material. It is found extensively in Great Britain, near the summit of the coal measures. In our country the black band ores are also associated with the coal measures, both in the anthracite and bituminous regions.

CHROMIC IRON OR CHROMITE, generally with 49.90 to 60.04 per cent. of chromic oxide, 18.42 to 35.68 per cent. of ferrous oxide, 10 to 12 per cent. alumina, 5.36 to 15 per cent. magnesia, and 4 to 6 per cent. silica, occurs usually massive, mixed with other iron ores or in serpentine. It is of an iron-black to brownish-black color and a faintly metallic lustre. Streak or powder, dark-brown. Fracture, irregular; specific gravity, 4.4 to 4.6; hardness 5.5, is not scratched by a knife. With borax bead it gives the characteristic indications of chromium. It is largely used in the preparation of chromium colors.

The following iron ores are not used for the making of iron and steel, but may nevertheless prove of value.

Iron Pyrites, usually in cubes and allied forms, sides often marked by fine parallel lines. Occurs also massive and contains 46.7 per cent. of iron and 53.3 per cent. of sulphur. Color, brass yellow; lustre, metallic; streak, brownish-black; fracture irregular; specific gravity 4.8 to 5.1; hardness 6 to 6.5; cannot be scratched with a knife, but is scratched by quartz, and scratches glass with great facility. Before the blow-pipe it burns with a blue flame, giving off an odor of sulphur, and ultimately fuses into a black magnetic globule. It is found in great abundance, and is used as a source of sulphur. It is easily distinguished from copper pyrites by its hardness, the latter being readily cut with a knife. From gold it is distinguished by its hardness and in not being malleable, and in giving off sulphurous odors in the blow-pipe flame.

Arsenical Pyrites or Mispickel contains 34.4 per cent. of iron, 19.6 per cent. of arsenic, and 46.0 per cent. of sulphur. It occurs in flattened prisms and also massive. Color, white; lustre, metallic; streak, gray; fracture, uneven; specific gravity 6.0 to 6.3; hardness 5.5; cannot be scratched with a knife, but is scratched by quartz. Heated before the blow-pipe it gives off white arsenical fumes of a garlic odor, and finally fuses into a black globule. It is abundant in mining districts, and sometimes is auriferous. With the improved processes now in use, it is possible to extract the gold profitably, and hence mispickel ores should be examined for gold.

THE GEOLOGY of the iron ores varies and may be divided into that of the magnetites, which are always derived from the granites, gneiss, schist rocks, clay slates, and, rarely, the metamorphic limestones.

The red hematites seem to be only an alteration derived from the magnetites, and belong to the same more ancient rocks as the latter.

The brown hematites (limonites) are derived from both the former and are generally sedimentary.

Very frequently in extensive magnetic regions, where the country back is mountainous, the brown ore has been formed in basins and knees and interlocked portions of the lower country, where ages of rains, storms and freshets have gradually transported and altered the magnetic ores of the upper regions and brought down these iron oxides to the

FIG. 53.

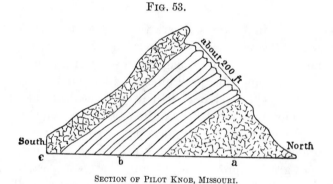

SECTION OF PILOT KNOB, MISSOURI.

a, Quartzite or siliceous rock. *b*, Red hematite iron ore alternating with siliceous matter. *c*, Siliceous rocks.

low lands, where they have been arrested and settled down in beds of brown hematite. This seems to have been the history of all the hematitic limonite beds and deposits; they are on the lower levels when they were formed, although in after ages they may have been uplifted.

Iron ores are, therefore, to be found in three general geologic regions: (1) in the earliest rocks; (2)

in the carboniferous, and (3) in the more recent or
sedimentary rocks, and in accordance with their
composition as magnetites and specular ores, as
carbonaceous or black band and spathic ores, or as
brown ores of the limonite order.

One of the most peculiar geologic conditions is
found in the Pilot Knob Mountain, wherein the
iron strata have been thrown up as in Fig. 53.

THE USE OF THE MAGNETIC NEEDLE IN PROSPECTING
FOR IRON.

In ordinary cases, where the surface is covered
with loose earth, it is common to search for magnetic iron ore with a magnetic needle or a miner's
compass, and for preliminary examinations it is
now the chief reliance. In using this instrument
considerable practice is required ; but this joined to
good judgment gives indications of the presence of
ore which are almost infallible. There has been
very great improvement, within a few years past, in
the methods of searching for magnetic ore as well
as in the instruments to be used for that purpose,
and the work is now well done by many persons.

In the Annual Report of the State Geologist of
New Jersey for 1879, W. H. Scranton, M. E., makes
a report, accompanied by a map, upon a magnetic
survey made at Oxford, Warren Co., New Jersey, to
determine the location of a vein, and the proper
places to sink shafts. Mr. Scranton finds Gurley's
Norwegian compass the best, though the slowest to
work with. He sums up the indications from the

magnetic needle in searching for ore, as it usually occurs in New Jersey, as follows:

" An attraction which is confined to a very small spot and is lost in passing a few feet from it, is most likely to be caused by a boulder of ore or particles of magnetite in the rock.

" An attraction which continues on steadily in the direction of the strike of the rock for a distance of many feet or rods, indicates a vein of ore; and if it is positive and strongest towards the southwest, it is reasonable to conclude that the vein begins with the attraction there. If the attraction diminishes in going northeast, and finally dies out without becoming negative, it indicates that the vein has continued on without break or ending until too far off to move the compass needle. If, on passing towards the northeast, along the line of attraction, the south pole is drawn down, it indicates the end of the vein or an offset. If, on continuing further still in the same direction, positive attraction is found, it shows that the vein is not ended; but if no attraction is shown, there is no indication as to the further continuance of the ore.

" In crossing veins of ore from southeast to northwest, when the dip of the rock and ore is as usual to the southeast, positive attraction is first observed to come on gradually, as the ore is nearer and nearer to the surface, and the northwest edge of the vein is indicated by the needle suddenly showing negative attraction just at the point of passing off it. This change of attraction will be less marked as the depth of the vein is greater, or as the strike is nearer north and south. The steadiness and continuance of the attraction is a much better indication of ore than the strength or amount of attraction is. The ore may vary in its susceptibility to the magnetic influence from impurities in its substance; it does vary according to the position in which it lies— that is, according to its dip and strike; and it also varies very much according to its distance beneath the surface.

"*Method of Using the Compass in Searching for Ore.* —It is sufficient to say that the first examinations are made by passing over the ground with the compass in a northwest and southeast direction, at intervals of a few rods, until indications of ore are found. Then the ground should be examined more carefully by crossing the line of attraction at intervals of a few feet, and marking the points upon which observations have been made, and recording the amount of attraction. Observations with the ordinary compass should be made and the variation of the horizontal needle be noted. In this way material may soon be accumulated for staking out the line of attraction, or for constructing a map for study and reference.

" After sufficient exploration with the magnetic needle, it still remains to prove the value of the vein by uncovering the ore, examining its quality, measuring the size of the vein, and estimating the cost of mining and marketing it. Uncovering should first be done in trenches dug across the line of attraction, and carried quite down to the rock. When the ore is in this way proved to be of value, regular mining operations may begin.

" In places where there are offsets in the ore, or where it has been subject to bends, folds, or other irregularities, so that the miner is at fault in what direction to proceed, explorations may be made with the diamond drill."

CHAPTER XI.

MERCURY OR QUICKSILVER. *Native mercury* in a pure state is rarely found. It occurs disseminated in liquid globules in cavities in cinnabar-bearing rocks, especially at or near the surface. It is bright, white, and of specific gravity 13.6 at 32° F.

Mercury is most commonly found in association with sulphur. Antimony is also a frequent companion, but not in chemical union. The ore of greatest industrial importance is

CINNABAR, or sulphide of mercury, found massive, of a granular texture, reddish color, and scarlet-red streak. Composition: Mercury 86.2, sulphur 13.8, when pure. It is the only regular and most valuable ore of mercury.

Hardness 2 to 2.5 ; specific gravity 8.99 ; sectile. Easily scratched with a knife, affording a deep red streak. Before the blow-pipe on charcoal it is volatile if pure, gives sulphurous flames if heated in an open tube, and mercury condenses on the sides of the tube, so that it may easily be seen with a lens or even the naked eye.

There is also a black sulphide, called *metacinna-*

(176)

barite, found in one locality in California; and, in California and Mexico, a sulphoselenide named *guadalcazarite* (81½ per cent. mercury, 10 sulphur, 6½ selenium) is sometimes encountered.

NATIVE AMALGAM. This is a mixture of silver and mercury, and when pure contains from 64 to 72 per cent. mercury. Color, silver-white; hardness 3–3.5; specific gravity 10.5–14. On charcoal before the blow-pipe, the quicksilver evaporates and the silver remains.

The quicksilver deposits at Almaden, in Spain, have a far remote history, for in the time of Pliny 10,000 lbs. were sent annually to Rome from these mines. They occur in upper Silurian slates, sometimes interstratified with beds of limestone; but the ordinary slates themselves, which are much contorted, rarely contain cinnabar. The enclosing rock usually consists of black carbonaceous slates and quartzites alternating with schists and fine grained sandstones.

At Idria, Austria, cinnabar is found in impregnated beds and stockworks, in bituminous shales, dolomitic sandstones and limestone breccias of triassic age, dipping 30° to 40°, and covered by carboniferous sandstones and shales in a reversed position. This deposit has been worked for nearly 400 years, and is said to become richer as the depth increases.

The quicksilver-bearing belt of California extends along the coast range for a distance of about 200 miles. According to a report by M. G. Rolland,

12

these deposits are generally impregnations in the cretaceous and tertiary formations. They seem to be richer when the beds are more schistose and transmuted. They are more or less closely in relation with serpentines, which are themselves sometimes impregnated with oxide of iron, sometimes in quartzose schists, in sandstones, more rarely in limestone rocks, limestone breccias, etc. Native mercury is found in some magnesian rocks near the surface. There are no defined fissures nor veins proper. The cinnabar with quartz, pyrites, and bituminous substances is sometimes disseminated in the rock in fine particles and spots, sometimes forms certain kinds of stockworks or reticulated veins and nests. The parts thus impregnated congregate and form rich zones, the size of which occasionally reaches 80 fathoms, and the percentage 35 per cent., and flat-like veins or lenticular deposits, the strike and dip of which agree with those of the schists of the country generally. These rich zones without defined limits gradually merge into poor stuff containing half a per cent., or more traces, and are of no value.

Sulphur Bank, one of the principal mines, was originally worked as a sulphur deposit. Sulphur in workable quantities is known to exist in some volcanic countries, and volcanic rocks are abundant at the California cinnabar mines.

BISMUTH. This metal occurs native, of a reddish silver-white color. Brittle when cold : hardness 2–2.5; specific gravity 9.7. Sectile when heated.

It carries, sometimes, traces of arsenic, sulphur, tellurium and iron.   On charcoal before the blow-pipe, it fuses and entirely volatilizes, leaving a coating which is orange-yellow while hot and lemon-yellow on cooling (this is the trioxide of bismuth).   It dissolves in nitric acid, but subsequent dilution causes a white precipitate.

Very little bismuth has been found in our country.   The metal occurs on the Continent of Europe, associated with silver and cobalt, also with copper ores.   Although there is but little call for it in the arts, a deposit or lode of bismuth would be valuable.

Where it has been found in the United States it has been associated with wolfram (tungstate of iron and manganese), also with tungstate of lime, with galena and zinc blende in quartz.

Its GEOLOGY is the same as that of copper ; it occurs in veins, in gneiss and other crystalline rocks.

NICKEL.   It does not occur native, except in meteorites.

Under the blow-pipe, nickel requires care and some practice.   On charcoal, with soda in the inner flame, it gives a gray metallic powder, attractable by the magnet.   In the borax bead in the outer flame it gives a hyacinth-red to violet-brown while hot, a yellowish or yellow-red when cold.   In the reducing or inner flame, a gray appearance is given.   These appearances are modified by the impurities and the amount of nickel in the mineral.   The wet process is the only method of determining the true value of a nickel-bearing mineral.

Its chief ores are:

SMALTITE, which is a combination of cobalt, iron and nickel, and arsenic in varying proportions. It will be more fully referred to, later on, under COBALT.

NICKEL ARSENIDE, "*copper nickel*," mineralogical name, *nicolite*. Composition: nickel 44.1; arsenic 55.9. It looks somewhat like pale copper, but contains no copper. Hardness 5–5.5, specific gravity, 6.67–7.33; streak, pale brownish to black; brittle. It frequently contains a little iron, and sometimes a trace of antimony, lead and cobalt.

If carefully treated under the blow-pipe with borax, it will show the iron if present, in the bead, and the cobalt and nickel by successive oxidations (*page* 179). But the nickel requires especial treatment, the detection of which we will speak of in this chapter.

There is another mineral, not properly an ore, called:

EMERALD-NICKEL, a carbonate of nickel, containing 28.6 water when pure. It forms incrustations on other minerals, like another called

MILLERITE, a sulphide of nickel forming tufts of very fine acicular, brassy-looking crystals, in cavities of the red hematite of Sterling Iron Mines in Northern New York, and velvety incrustations on ores in Lancaster Co., Penna., where nickel is found and worked. In the former place no nickel abounds, but in the latter it has been found in paying quantities. But the sulphide forms at the latter place vary very much, as examined under the microscope, from the acicular crystals found in the ores at Sterling, N. Y., and yet they are of the same chemical combination. The ore upon which the tufts of velvety covering are found at the Gap Mine, Lancaster Co., Penn., is pyrrhotite or sulphide of iron, holding 4 to 5.9 per cent. nickel in composition; that of Sterling, N. Y., is the red hematite.

The sources of nickel discovered in Sudbury, Canada, north of Georgian Bay, yield nickel in *pyrrhotite* (sulphide of iron), and apparently also in *chalcopyrite*, whose typical composition is copper 34.6, iron 30.5, sulphur 34.9. It is a mineral of brass-yellow appearance, and one which furnishes the copper of commerce at the Cornwall Mines (Eng.) and at the copper beds in Fahlun, Sweden. In the latter place it is imbedded, as it appears to be in the region of the Sudbury Mines, only that the Sudbury ore is imbedded in pyrrhotite and the Swedish in gneiss.

The chalcopyrite does not mix intimately with the nickel ore so as to form a homogeneous mass; it occurs by itself in pockets or threads, etc., but inclosed with massive pyrrhotite, which, while it may have more than 30 per cent. of nickel present, does not show any sign of the changed composition.*

This per cent. is far above the average of nickel in the pyrrhotite, which seldom carries less than $2\frac{1}{2}$ per cent. or more than 9 per cent. of nickel.

* Dr. E. B. Peters, Manager of the Canada Copper Company.

The following new ores of nickel are reported by Dr. Emmons from Sudbury, Canada :

Foleyrite, of a bronze-yellow color, grayish-black streak, and metallic lustre. It occurs massive and contains 32.87 per cent. of nickel. Its specific gravity is 4.73, hardness 3.5.

Whartonite contains 6.10 per cent. of nickel. It has a pale bronze-yellow color, black streak and metallic lustre. Specific gravity about 3.73 ; hardness about 4.

JACK'S TIN OR BLUEITE contains 3.5 per cent. of nickel. It is of an olive-gray to bronze color, metallic lustre and black streak. Specific gravity 4.2 ; hardness 3 to 3.5.

ANALYSIS OF ORES FOR NICKEL AND COBALT.

As this analysis requires care, we give the following method in full :

1. Reduce finely 50 grains of the ore. Put it in a dry beaker-glass and pour over it a mixture of one part sulphuric acid with three parts nitric acid, both pure and concentrated, or 40 to 50 c.c. to 2 grams of ore.

2. Heat the covered beaker on a sand-bath to near 212° Fah. for two hours. Then partly uncover, and evaporate the nitric acid entirely.

3. Cool and add 100 or more c.c. of water and let it stand for four hours ; the insoluble residue is LEAD sulphate, silex, etc.

4. Filter off the soluble part and place the moist lead sulphate in a beaker and dissolve it by first

pouring in ammonia (20–25 c.c.), and next acetic
acid till it is decidedly acid. The sulphate now
dissolves if kept warm for some twenty minutes;
Filter and wash, aud if any residue remains (silex,
etc.), reserve for future examination.

5. The LEAD is now separate, but if the amount
is sought, pass a current of hydrogen sulphide
through the solution till the lead is entirely precipitated. Filter, dry, place the residue in a porcelain crucible and heat to a low-red heat, passing a
current of dry hyrogen into the crucible while
heating to prevent any oxidizing of the sulphide.
When the crucible and contents remain the same
in weight, the last weight of the lead sulphide is the
correct amount. Of this weight, 86.61 parts in 100
are lead, 13.39 are sulphur.

If the ore has no lead in it, the above work is
omitted entirely. The likelihood of lead may be
tested qualitatively from a small quantity dissolved,
precipitated by hydrogen sulphide, and the precipitate determined by the blow-pipe on charcoal giving
the lead coating, and with soda, the metallic globule.

6. To SEPARATE THE COPPER. The filtrate remaining after the insoluble lead sulphate was filtered off, as in No. 4, now contains whatever the
mineral is composed of, copper, iron, nickel, cobalt,
etc. Dilute the filtrate to about 500 c.c., heat to
nearly boiling, and pass hydrogen sulphide through
it, and thus precipitate all the copper after adding
1 or 2 c.c. of hydrochloric acid. Filter, wash, dry,
and ignite the precipitate in an atmosphere of

hydrogen.   The result will be pure Cu_2S, from which the copper may be ascertained as 79.85 parts of the whole weight of Cu_2S.

7. Concentrate by evaporization the filtrate of No. 6 remaining after the copper was separated, add 1 or 2 c.c. of nitric acid, and boil the filtrate two or three minutes, let it become nearly cold, add an excess of ammonia, and let it stand in a warm place half an hour.

8. Filter the precipitate into a porcelain dish and redissolve the iron oxide (hydroxide) with hydrochloric acid poured slowly into the filter, complete washing of the filter with hot water, reduce the free acid in the filtrate with ammonia, then very nearly neutralize it carefully with sodium (metallic) or ammonium carbonate ; the solution must remain clear, though dark red, if much iron is present. Now add a strong neutral solution of ammonium or sodium acetate (not in large excess), and then boil a short time.   When rightly performed the iron oxide precipitate will settle rapidly, and the supernatant liquor will be clear.   Wash rapidly with boiling water, and, at first, separate the clear part by decantation, and then filter.   If great exactitude is required, redissolve in hydrochloric acid, and once more precipitate with the acetate just as before. Add this filtrate to the ammoniacal filtrate mentioned at the beginning of No. 7 paragraph.

The iron is now separated as basic ferric acetate, and it is almost, if not entirely, separated from all nickel and cobalt which are yet in solution.

9. The first filtrate, no 7, contains all the nickel and cobalt. It must now be concentrated to about 250 c.c. If it is slightly acid, proceed ; if not, then add muriatic acid until it is very slightly acid. Now heat the filtrate in a beaker to gentle boiling, and pass hydrogen sulphide through the liquid. A black precipitate follows ; if nickel sulphide with cobalt sulphide, they are together.

10. Filter, wash, and dry ; incinerate the filter-paper with the precipitate if very small in quantity, otherwise separately ; heat in porcelain crucible. Dissolve in aqua regia (nitro-muriatic acid), and treat it till only yellow sulphur remains, evaporate and expose the residue to a heat of 180° Fah. to make any silica insoluble. Moisten with a few drops of muriatic acid, add 20 c.c. of water to dissolve the salts, add some solution of hydrogen sulphide to separate any copper or lead which may have escaped separation, filter into a porcelain dish and concentrate all to about 100 c.c.

11. Boil gently, and while boiling add *pure* sodium carbonate solution until the liquid is slightly alkaline. Continue boiling a few minutes, add a few grains of pure soda solution (sodium hydroxide). This is best prepared freshly by dropping a small ball of metallic sodium into a half ounce of water in a platinum dish or crucible, or, not so well, in a porcelain dish. Heat to a boiling again a few minutes till all the nickel and cobalt are precipitated, wash the precipitate with boiling hot water by decantation, and finally on the filter, until a drop on

polished platinum shows no residue.   After drying
the precipitate remove it to a piece of glazed paper;
cover with a bell-glass.   Then incinerate the filter
till the carbon has entirely disappeared, add it to
the precipitate already obtained, place all in a crucible, cover it and expose to heat to redness, and,
finally, if desired, reduce the oxides to the metallic
condition by ignition under a stream of hydrogen.

12. As this process of reduction to metal is sometimes very useful, we give a simple plan of apparatus for this purpose.   Get a half-pint, wide-
mouthed pickle bottle and introduce two glass tubes
of a quarter inch diameter into a cork fitting the
mouth, after having nicely adjusted the cork to the
mouth of the bottle.   The tubes may be easily bent
and blown as in A B, Fig. 54, over the flame
of an alcohol lamp, before permanently fastening
them in place.   To blow a funnel end, heat the end
of the tube to softness and mash it together, her-
metically seal, then reheat rapidly, roll it between
finger and thumb while gently blowing at the other
end until swollen large enough, then, with pincers,
break it or chip it off; if enlarged twice or three
times the diameter, it is large enough for the purpose.   The tubes intended to be bent should be
rapidly rotated in the enlarged flame until red-hot,
and then bent to the right angle and gradually
cooled.

It is well to make another of these bottles for drying the hydrogen, as in B.   Introduce the tube as
shown in the figure, wherein B represents the drying

bottle in which is placed a quantity of fragments of chloride of calcium of the size of peas or even smaller. In putting the cork with tubes into this bottle, the bottle should be on its side and rolled while introducing the longer tube into the calcium chloride, so that the fragments may not obstruct the tube as it is pushed down. The exit tube may be bent or straight, and properly sized india-rubber tubing may be fitted over the ends so as to make connections. A common clay stem smoking pipe

Fig. 54.

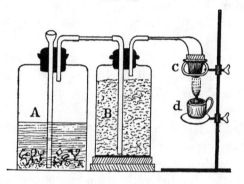

arranged as in the figure, with the bowl inverted into the crucible which is placed on a wire support on a retort stand, c, is quite sufficient. The usual alcohol blast lamp, d, is necessary for this operation. To put the apparatus to work it is only necessary to introduce some three or four ounces of broken-up pieces of zinc into A, together with water sufficient to half fill the bottle, cork up with the tubes arranged as above, and pour into the funnel-shaped

tube common oil of vitriol gradually, until the gas
begins to come over, then stop as the water becomes
heated, and the gas will increase without addition.
You may now prepare your crucible, and, when in
place, and the tubes all arranged, the gas may be
made to come over more rapidly by adding a little
more oil of vitriol drop by drop.

13. The crucible should be weighed after cooling
and replaced, the flame of the blast lamp relighted,
and red heat renewed under the hydrogen apparatus
until the crucible, when again weighed, shows no
alteration in weight. The oxide has now been reduced to the pure metal form, and it may then be
cooled.

In the case of the analysis we are now upon, the
metallic reduction will be that of both nickel and
cobalt, and they will appear as a dark powder in
the bottom of the crucible.

When the hydrogen apparatus is no longer to be
used, the generator bottle A should be washed thoroughly and the zinc also ; the latter may be left in
the bottle and the cork replaced loosely, but the
cork must be removed from bottle B, and a tight-
fitting cork be used in its place, as the chloride may
be used again. All is ready for another operation
by simply replacing and adding water and acid as
before.

14. SEPARATION OF NICKEL AND COBALT. The
two metals should be weighed in order that if the
cobalt be found, the nickel may be known by the
difference. Dissolve the two metals in nitric acid

and evaporate them till there is no free nitric acid.
Next add about 6 to 8 grams (100 grains), potassium nitrate dissolved in 10 to 15 c.c. of hot water.
If any flocculent particles appear, add a little
acetic acid, just sufficient to dissolve them, and
now a precipitate of cobalt (as tripotassium cobaltic
nitrite), takes place slowly. The whole volume
should now be 15 to 20 c.c. Cover the beaker containing it with glass, and set it aside in a warm
place for twenty-four hours. Filter, wash with a
solution of potassium acetate (which may be made
by neutralizing acetic acid with crystallized potassium bicarbonate, leaving the solution slightly
acid), and proceed to more efficiently separate the
cobalt as a metal, as follows :

Dilute the filtrate, heat, and precipitate with
caustic soda (sodium hydroxide), wash the greater
part of the saline matter out and then dissolve the
precipitate in nitric acid, evaporate to dryness, add
two or three drops of nitric acid and dissolve in a
small volume of water, filter, concentrate the filtrate, and repeat the process of separation of potassium nitrite as before. Put this precipitate, with
the filter-paper, into a beaker, add about 100 c.c. of
water, heat, add muriatic acid to dissolve it, separate
the filter-paper by filtering it and washing it in a
funnel, evaporate the solution on a water-bath, and
let it remain on the water-bath two or three hours
to render the silica insoluble, then moisten with
muriatic acid, add water, filter, and convert the cobalt to metallic form, as was done before for both

nickel and cobalt, namely, as in paragraph No. 11. The cobalt is now entirely separate from the nickel. Weigh it, and by difference from the weight of the two determine the weight of nickel as suggested in No. 14. The amount of nickel is now known by weight, and readily compared with the whole amount of the original weight of ore employed at the beginning.

If the above process is carefully followed out, in a mineral containing lead, copper, iron, cobalt, and nickel, the cobalt and nickel are separated with great exactness.

But the main ore of nickel is pyrrhotite, and, as in the Gap Mine, Lancaster Co., Penn., and in the Sudbury Mines, Canada, pyrrhotite contains only iron and nickel, seldom cobalt enough to notice. So that much less work is required, as follows: Pulverize, dissolve in muriatic acid in a flask. If much free acid is present, nearly neutralize with sodium or ammonium carbonate; the solution should be clear, but, if there is much ferric chloride, it should be of a deep-red color; now do as directed in No. 8, to add the ammonium acetate, and proceed as before.

In view of the importance of nickel-steel armor plates, prospecting for nickel is a work of unusual interest. In addition to the discovery of the nickel pyrrhotite in Canada, which we have already noticed, new discoveries have been reported from New Caledonia, an island 900 miles east of Australia. The ore is a nickel silicate and has been named

GARNIERITE, after M. Garnier, its discoverer. It is also found in Oregon. It contains from 8 to 10 per cent. of nickel, has a green color and yields an un-colored streak.

The mines at the Gap, Lancaster Co., Penna., are considered nearly, if not quite exhausted. There is now, as may readily be imagined, increased demand for nickel ores.

COBALT.—Cobalt does not occur in native form. The following are the minerals of importance:

SMALTITE seems to be composed of cobalt, nickel, iron and arsenic; the typical form is arsenic 72.1, cobalt 9.4, nickel 9.5, iron 9 = 100. Hardness 5.5–6; specific gravity 6.4–7.2. Color, tin-white, sometimes iridescent. Streak, grayish-black. Brittle. Before the blow-pipe, on charcoal with soda, the arsenious acid fumes are given off, and the garlic smell is plainly observed. With borax for the bead the assay may be made to show (with successive heatings), the reactions first of iron, then cobalt, and nickel, provided the operator is skillful in oxidizing the powdered ore by cautious degrees; when one borax bead shows iron reaction by a certain amount of carefully applied $O\ F$ to the bead, try another with increased degree of oxidization until you per-ceive the cobalt blue and nickel brown, if both are present.

COBALTITE is composed of sulphur, arsenic, and cobalt in the typical proportions of 19.3, 45.2, 35.5 = 100, but it frequently, as a mineral, contains iron. Hardness 5.5; specific gravity 6–6.3. Under the blow-pipe, in an open tube, it sends off sulphurous fumes and a sublimate of arsenous acid. With borax bead gives the blue of cobalt. Dissolves in warm nitric acid, separating the sulphur and arsenic.

Cobaltite and smaltite are valuable as affording the greater part of smalt of commerce, and the'former is used in porcelain painting.

ERYTHRITE is a soft (1.5–2.5), peach-red mineral of specific gravity 2.9, transparent or translucent, sometimes pearl- or greenish-gray.

Composition, typical, arsenic 38.43, cobalt oxide 37.55, water 24.02 = 100.

In a closed tube, under blow-pipe, it yields water and turns bluish. Gives the usual blue for cobalt in the borax bead.

Valuable for the manufacture of smalt. It is sometimes known as "*cobalt bloom.*"

LINNÆITE. This is valuable for the large amount of both cobalt and nickel it sometimes contains. Hardness 5.5; specific gravity 4.8–5; metallic lustre ; color, pale steel-gray, tarnishing to red. Composition, sulphur 42, cobalt 58 = 100. but cobalt is replaced by large amounts of nickel, and sometimes copper. Some specimens from Mineral Hill, Maryland, and from Missouri, have yielded as high as 29.56 and 30 per cent. nickel, with 21 to 25 per cent. cobalt in the same specimen, but with a small amount of iron (3 per cent.).

EARTHY COBALT, OR COBALT WAD (*Asbolite* is the mineralogical name), occurs as a bog ore, with manganese, iron and copper, and nickel. It is blue-

black at times, has a hardness of 1 to 1.5, and
specific gravity of 2.2 to 2.6. It sometimes contains
up to 35 per cent. of cobalt oxide.

The geological position of cobalt is in the earlier
rocks, as the chlorite slates with chalcopyrite, blende,
and pyrite, as in Maryland. Sometimes the ore is
found in cavities in the limestone of the carboniferous age, as in Great Britain. The tin-white cobalt
is found in the gneissic and primitive rocks, as in
Norway. Linnæite is found at Mine la Motte, Mo.,
in masses, sometimes in octahedral crystals among
its rich ores of lead and nickel.

CADMIUM. Of this mineral but one ore is known,
namely, the sulphide, or GREENOCKITE, with 77.7
per cent. cadmium. Color, honey to orange-yellow
and brick-red; in hexagonal prisms; hardness 3 to
3.5; specific gravity 4.5 to 4.908. Before the blow-
pipe, on charcoal with soda, it yields a red-brown
deposit. Cadmium is frequently associated with
zinc ores, some varieties of sphalerite or blende containing 3.4 per cent.

Metallic cadmium is white like tin, and shares
with it the property of emitting a crackling sound
when bent. It is so soft that it leaves a mark upon
paper.

13

CHAPTER XII.

ALUMINIUM. The distribution of aluminium in nature is very wide, rivaling that of iron, yet there are but few minerals which serve as sources of the metal. These are: *Bauxite*, a limonite, in which most of the iron is replaced by aluminium; soft and granular, with 50 to 75 per cent. alumina. *Corundum*, crystalline and very hard, specific gravity 4, generally quite pure, but too valuable for abrasive purposes to be used as an ore. *Diaspore*, hard and crystalline, specific gravity 3.4, with 64 to 85 per cent. alumina, and ordinarily quite pure. *Gibbsite*, stalactic, specific gravity 2.4, containing, when pure, 65 per cent. alumina. *Aluminite*, specific gravity 1.66, a sulphate of aluminium found in large beds, chiefly along the Gila River, in New Mexico, containing about 30 per cent. alumina, and easily soluble in water. *Cryolite*, specific gravity 2.9, easily fusible, and when fused its specific gravity is about 2. It contains 40 per cent. aluminium fluoride and 60 per cent. sodium fluoride. All clays contain a large percentage of aluminium, but always in the state of silicate, and the difficulty of removing this silica has so far prevented the employment of clay as an ore of aluminium.

Of the ores above named the most important is

BAUXITE, of which there are vast deposits at Baux, near Arles, in France, in Ireland, and in Alabama, Arkansas, the Carolinas, Georgia, Tennessee and Virginia.

The Arkansas deposits are said to cover a large area, and to reach a thickness of 40 feet, forming an interbedded mass in ferruginous Tertiary sandstone.

The Alabama deposits are better known, and all occur in the lower part of the lower Silurian formation. The district has been badly broken up by sharp folds and great thrust faults, and the mineral occurs as pockets in close association with brown iron ore (limonite) and clay.

Bauxite has to undergo purification for the purpose of the aluminium manufacturer. Several methods are used :

1. It is chosen as free from iron as possible, and is roasted at a low red heat, and afterwards treated with sulphuric acid, which combines with the alumina present, forming sulphate of alumina. This is readily dissolved by water, leaving the great bulk of silica and iron behind. The solution of sulphate of alumina is allowed to settle, the supernatant liquid is siphoned off into an evaporating tank and evaporated to dryness. The dry sulphate of alumina is calcined at a red heat, driving off the sulphuric acid, leaving as a residue anhydrous alumina.

2. The bauxite is treated either by fusing with

carbonate of soda and dissolving in water, or by boiling it with a strong solution of caustic soda. In either case a solution of sodium aluminate is obtained, which is filtered from the residue of silica and ferric oxide, and decomposed into aluminium hydrate and carbonate of soda by pumping carbonic acid gas through it. After a thorough washing, the hydrate is calcined at a high heat, and the resulting alumina is finely ground.

The ore next in importance is

CRYOLITE, of which there is practically only one productive mine, that at Ivigtut, South Greenland. The mine is worked as a quarry, and has been opened 450 feet long, 150 feet wide and 100 feet deep, while diamond drills have proved the permanence of the ore for a further depth of 150 feet. The vein appears to widen with depth, but the quality of the mineral becomes inferior. About 10,000 tons of cryolite annually are shipped to the United States.

With the blow-pipe, on charcoal, cryolite fuses to a clear bead, becoming opaque on cooling. After long blowing with O F the assay spreads out, the fluoride of sodium sinks into the charcoal, and the suffocating odor of fluorine is given off and the aluminium remains as a crust which, if touched with a little cobalt solution and gently heated, gives a blue color of alumina. If some of the cryolite is powdered and placed near the open end of a glass tube and the flame from the blow-pipe turned carefully on it, the fluorine will be freed and will etch the glass, showing corrosion and proving the presence of fluorine. Besides as a source of the metal aluminium, cryolite is used as a flux, and largely for the manufacture of alumina of soda.

While the older processes of aluminium manufacture, dependent on the reduction of the double chloride of aluminium and sodium, must always have a scientific interest, they have been beaten out of the field of commercial industry by the newer electrolytic methods, of which there are four varieties. In England and America Cowles' and Hall's patent are followed; on the Continent, Heroult's and Minet's. They are all virtually modifications of the original Deville-Bunsen process, maintaining fusion by the heat of the electric current.

CORUNDUM AND EMERY. While corundum and emery are very nearly allied mineralogically, they are sharply distinguished in commerce. Corundum is almost a pure alumina, but emery is contaminated with a large proportion of iron oxide, ranging generally between 20 and 33 per cent. Physically they are also distinguished by the following features: Corundum is variously colored, commonly gray, but never black. It is much harder than emery, with sharper edges and cuts more deeply and rapidly. It is, however, more brittle and therefore less durable. Emery is practically always black.

Corundum is infusible before the blow-pipe, and is not affected by acids or by heat. It crystallizes in six-sided prisms, often irregularly shaped, and sometimes occurs in granular masses. Transparent or opaque. Lustre, glassy, sometimes pearly. Fracture uneven or conchoidal. Specific gravity, 3.9 to 4.2. Hardness 9, it being, next to diamond, the hardest of minerals. It is generally found associated with some member of the chlorite group, and a series of aluminous minerals in part produced from its alteration.

The blue variety of corundum is called *Sapphire*, the most esteemed shade being deep velvet blue; the blood red variety is the ORIENTAL RUBY, which can be readily distinguished from other red gems by its superior hardness; the bright yellow variety is the ORIENTAL TOPAZ, distinguished by its hardness from the topaz, yellow tourmaline and false topaz; the bright green is the ORIENTAL EMERALD; the bright violet, ORIENTAL AMETHYST. One variety exhibits a six-rayed star inside the prism, and is called the ASTERIAS. Ruby is the most highly prized form of this mineral.

Corundum has been found in a large number of localities in the United States, but only a few places have been actual producers. The emery vein or bed at Chester, Mass., has furnished a large quantity of the mineral, but the chief American source at present is a belt of serpentine that extends from southwestern North Carolina into Georgia. It is an altered olivine rock, and has gneiss for its immediate association, and along the contact of the two are found the veins or beds of decomposed rock which have the corundum disseminated through them. Corundum Hill, in North Carolina, and Laurel

Creek, in Georgia, are the chief producers. The mineral is crushed, sifted and washed, and thus comes to market in various sizes. Care is taken to avoid making undue amounts of the finest product, or " flour," for this has less value than the coarser grades. The chief European sources of emery are the Greek island of Naxos and Asiatic Turkey.

The usual test for the quality of a sample of emery or corundum is to compare a weighed sample with an equal amount of the standard grade or of some well-recognized brand. Two weighed pieces of plate glass of convenient size are then rubbed together with the sample between, and the process is continued until the grit has disappeared and the plates no longer lose in weight from the abrasion. The amount of loss is a measure of the hardness and abrading power of the sample, the better grade giving the greater loss.

ANTIMONY. This metal occurs in three forms, namely, the oxide, *senarmontite*, containing 83.56 per cent. antimony; the sulphide, *stibnite, antimonite* or *antimony glance*, affording 71.8 per cent., a sulphoxide, *kermesite*, giving 75.72 per cent., in addition to some unimportant combinations with silver, etc. While it may be said that antimony is somewhat widely distributed in nature, yet, owing to cost and difficulties in extraction, only comparatively few mines affording a rich ore can be profitably worked. Beyond the considerable quantities of oxide coming from Algiers and of kermesite from Tuscany, almost the entire output is in the form of

STIBNITE, which contains 78.8 per cent. antimony and 28.2 sulphur. Hardness 2; specific gravity 4.5 ; metallic lustre ; color and streak, lead-gray; sectile. When pure, perfectly soluble in muriatic acid.

Before the blow-pipe, on charcoal, it fuses, spreads out, gives sulphurous and antimonious fumes, coats the charcoal with white oxide of antimony ; this coat, treated in $R\ F$, tinges the flame greenish blue.

Foremost in antimony production stands Portugal, due principally to the mining district of Oporto. The geological formations of Portugal are chiefly igneous and old sedimentary. The most favorable rocks for good antimony ore are bluish gray argillaceous Silurian shales.

Among the other European centers of production, the Bohemian mines are in granite and mica schist ; the Hungarian in granite—sometimes auriferous ; the Styrian in dolomite, and the Turkish also in granite. Victoria, New South Wales, and Western Australia are large producers of auriferous stibnite. In New Brunswick, antimony is mined in a quartz and calcite gangue in clay-slates and sandstones of Cambro-Silurian age.

Within the United States stibnite has been found in a number of places, all in the West. At San Emigdio, Kern Co., California, it is contained, with quartz gangue, in a vein in granite. The vein varies in thickness from a few inches to several feet. Several other small deposits occur in San Benito Co. and elsewhere in California. Stibnite has also been discovered in Humboldt Co., Nevada, and in

Louder Co., not far from Austin, in a quartz gangue. Some remarkable deposits occur in Iron County, Utah, as masses of radiating needles, which follow the stratification planes of sandstone and fill the interstices of a conglomerate. Stibnite˙ is found in Sevier Co., Arkansas, filling veins, with a quartz gangue, in sandstone.

MANGANESE. The ores of manganese are divided into three general classes :

1. Manganese ores.
2. Manganiferous iron ores.
3. Argentiferous manganese ores.

WAD is the name given to manganese oxide. It is found in earthy compact masses of a dark brown color, chiefly oxide of manganese and water.

Easily recognized under the blow-pipe, as it gives (in minute quantities), in the borax bead, a violet color in the O F, but disappears when the R F is turned upon it, and reappears when the O F is repeated.

It is found in beds varying from several inches to a foot or more in thickness. Hardness 1 to 3 ; specific gravity 2.3 to 3.7. Wad is used as a flux in iron smelting, and in a lixiviated state as a paint.

PYROLUSITE. This is the peroxide or dioxide, with 63.2 per cent. of manganese and 36.8 per cent. oxygen. Its crystalline form is the rhombic prism and it generally occurs in the form of minute crystals grouped together and radiating from a common centre. It has an iron-black or steel-gray color, a semi-metallic lustre and yields a black streak.

Specific gravity 4.7 to 5; hardness 1.5 to 2.5; infusible before the blow-pipe, and acquires a red-brown color. On heating it generally yields some water and loses 12 per cent. of oxygen. With borax, soda and microcosmic salt it shows manganese reaction. It dissolves in hydrochloric acid, when heated, with vigorous evolution of hydrogen.

PSILOMELANE occurs massive, frequently shelly, seldom fibrous; color, iron-black to bluish-black, streak bluish-black and shining; fracture, conchoidal to smooth. Specific gravity 4.1 to 4.2, hardness 5.5 to 6. Before the blow-pipe it yields manganic oxide, giving off oxygen. It is soluble in hydrochloric acid, chlorine being evolved. The powdered ore colors sulphuric acid red. Psilomelane contains from 40 to 50 per cent. of manganese, and some baryta and potassa. A solution in hydrochloric acid, of the variety containing baryta gives a heavy white precipitate with sulphuric acid.

MANGANESE CARBONATE (*Rhodochrosite* is the mineralogical name) occurs in spherical and nodular aggregations of cauliform texture or in compact masses of granular texture. It is rose-red to raspberry-red in color, by weathering frequently brownish, with a glassy or mother-of-pearl lustre. It cleaves like calcite. It contains 61.4 per cent. of manganese protoxide and 38.6 per cent. of carbonic acid, with part of manganese frequently replaced by calcium, magnesium, or iron. Specific gravity 3.3 to 3.6; hardness 3.5 to 4.5. Before the blow-pipe it is infusible and becomes black. From similar minerals it is distinguished by its rose-color and the manganese reaction with soda and borax ; and from silicate of manganese by its inferior hardness, its effervescence with acids and its non-fusibility.

The manganese in ores of the third class is valuable, even where the silver alone is sought, as it facilitates the work whereby the silver is extracted ; this it does because of its fluxing quality.

Virginia, Georgia and Arkansas are the chief producing States.

The geological position of manganese in some places seems to be the same as with the red hematite, as in Virginia.

In Tennessee it is found in the foot-hills of the mountains, four miles from Newport, Cocke Co., in pockets, and is a black oxide of 48 per cent. metallic manganese.

In Vermont it is found near a siliceous limestone, and in the vicinity of brown hematite ores. It exists in the triassic formation in Bosnia.

In North Carolina it is found in light-colored gneissic schists.

CHAPTER XIII.

Alum is hydrated sulphate of potash and alumina, and is best known by its astringent sweetish taste. Hardness 2 to 2.5. Specific gravity 1.8. Soluble in its own weight of boiling water. Found incrusting and impregnating dark slaty rocks, with yellow streaks. Used in dyeing and calico-painting, candle making, dressing skins, clarifying liquids, and in pharmacy.

Apatite, Phosphate of Lime, occurs in six-sided prisms, also in masses. It is transparent or opaque ; colorless, white, yellowish, green, violet, with a glassy lustre, and yields always a white streak. Fracture, conchoidal or uneven. Specific gravity 3.16 to 3.22 ; hardness 5. In thin laminæ it is fusible with difficulty before the blow-pipe ; when moistened with sulphuric acid tinges the flame greenish. It is soluble in hydrochloric and nitric acids without effervescence. From beryl it is distinguished by its inferior hardness and its solubility in acids. It occurs in rocks of various kinds, but more frequently in those of a metamorphic crystalline character, as in Laurentian gneiss, which is usually hornblendic, granitic or quartzose in char-

(204)

acter, in Canada, and in association with granular limestone. It is also found as an accessory mineral in metalliferous veins, especially those of tin, and beautifully crystallized and of various colors in many eruptive rocks. It also occurs in veins by itself, mostly in limestone, but sometimes in granites and schists. In these deposits apatite is also found as concretions, sometimes showing a radiated structure, but of an earthy appearance externally.

In sedimentary formations where a considerable accumulation of fossils has provided the phosphate of lime it occurs in two principal forms, namely *coprolites*, which are excreta of large animals, especially saurians, and concretions formed at the expense of the same coprolites, together with shells, bones, etc. The richest of these deposits are from Lower Cretaceous to Lower Jurassic in age, but phosphatic deposits are found and worked in sedimentary deposits of all ages.

The principal use of apatite is as a source of phosphoric acid and phosphorus, and before the discovery of the phosphate-rock deposits in Florida was largely sold to the manufacturers of fertilizers.

ARSENIC is found in the mineral kingdom partly in a metallic state, partly in combination with oxygen, sulphur and other bodies.

1. *Native Arsenic* occurs seldom distinctly crystallized, but usually in fine granular, spherical or nodular masses. Specific gravity 5.7 to 5.8; hardness 3.5; brittle; uneven and fine-grained fracture; metallic lustre; color, whitish lead-gray, usually

with a grayish-black tarnish; evolves an odor of garlic on breaking; contains occasionally more or less iron, cobalt, nickel, antimony and silver. Before the blow-pipe it quickly volatilizes before fusing, giving off white fumes having an odor of garlic. Native arsenic occurs especially in veins in crystalline slates and transition rocks in subordinate quantities associated with ores of silver, lead, cobalt and nickel.

2. *Realgar*, with 70.029 per cent. of arsenic and 29.971 per cent. sulphur. Color, red; crystallizes clinorhombic; fracture conchoidal to splintery; hardness 1.5 to 2.0; specific gravity 3.4 to 3.6. It is but slightly affected by acids; soluble with a deposit of sulphur in aqua regia, and in concentrated potash lye with separation of dark brown sulphuret of arsenic. From ruby silver and cinnabar, it is readily distinguished by its inferior hardness, slighter specific gravity and orange-yellow streak, the streak of the two above-mentioned minerals being cochineal-red.

3. *Orpiment*, with 60.9 per cent. of arsenic and 39.1 per cent. of sulphur; occurs in nature, but for industrial purposes is mostly artificially prepared. The mineral has a lustrous lemon-yellow or orange-yellow color, is cleavable into thin, flexible, transparent laminæ; hardness 1.5 to 2; specific gravity 3.4 to 3.5; soluble in nitric acid, potash lye and ammonia.

ASBESTUS. Fibrous. Color, green or white. The asbestus of commerce is practically a finely fibrous form of serpentine, that is to say, it is essentially a hydrated magnesium silicate. Every deposit of serpentine is a possible repository of asbestus. It occurs in seams half an inch to several inches in width, running parallel to or crossing one another, the width of each seam making the length of the fibre. Canada furnishes at present a large portion of the world's supply of asbestus. The profitable mining, however, is at present confined to a small area in the great serpentine belt of the Province of Quebec, that lies to the south of the St. Lawrence River. In the form of a rough cloth asbestus is used for covering steam-pipes, and for many purposes requiring an incombustible material.

BARYTES, or *barium sulphate*, commonly called *heavy spar*, occurs in tabular, glassy crystals, and also in dull masses in veins. of various rock formations. Color, white or tinted ; transparent or translucent ; lustre, vitreous or pearly. Specific gravity, 4.3 to 4.7. Hardness, 3 to 3.5. It is readily distinguished by its great comparative weight. When heated in the blow-pipe flame splinters fly off the crystals. It fuses with difficulty, and imparts a green tinge to the flame. After fusion with soda, it stains silver coin black. It is not acted upon by acids.

In the United States barytes is found in many places, it being mined in Virginia, Missouri, New Jersey and other states. It frequently occurs in connection with lead and zinc deposits forming the gangue of the metal-bearing vein. The best varieties of barytes are the white and gray. The chief use of barytes is as a pigment, as a cheaper substitute for white lead. It is also used as a make-weight by paper manufacturers, etc.

The carbonate of barium, *witherite*, is a much less common mineral than the sulphate. It sometimes occurs in crystals, but the more common form is that in fibrous masses. It occurs in veins. It fuses easily in the forceps, and gives a yellow-green flame. In hydrochloric acid it dissolves with effervescence, the solution yielding a heavy white precipitate (barium sulphate) if a little sulphuric acid is added. Witherite is used in the refining of sugar, and also in the manufacture of plate glass.

BORAX. Monoclinic. Fracture, conchoidal. Lustre, vitreous to resinous. Color, white, sometimes grayish, bluish, or greenish. Streak, white. Taste, slightly alkaline and sweetish. Translucent to opaque. Principal producing localities in the United States: the Columbus and Rhodes marshes in Nevada, the Saline marshes in California. In the Calico district the borate of lime is taken from a fissure vein, and this district is the only place in the world where deep mining for borax is carried on.

Borax is used in medicine and as an antiseptic by meat packers and others. Its chief use, however, is as a flux in metallurgical operations, in enameling, glazing of pottery and in the manufacture of glass.

CLAYS. The clays are all products of alteration from other minerals. Their composition is variable and they do not crystallize. The true clays are all plastic and refractory to a greater or less degree, and on these properties their value for industrial purposes depends. Pure *kaolin* is the type of all the clays.

The presence of alkalies in clays is objectionable, as it renders them fusible, as also do many other oxides. Iron is not only objectionable on the score of fusibility, but also as coloring matter. The presence of too large a proportion of water, carbonic acid or organic matter, causes clay to contract under the action of fire, and the same result will ensue if the clay is partially fusible.

The soft clays are divided into *kaolin, China clay, or porcelain clay,* which is nearly pure, and is derived from the decomposition of feldspar in pegmatite or granite; *plastic or pottery clay,* not so pure as kaolin, and *bole,* containing a large percentage of oxide of iron. *Fuller's earth* is a kind of clay composed, when pure, of 45 per cent. silica, 20 to 25 per cent. alumina, and water. It was formerly largely used as an absorbent in fulling or freeing woolen fabrics and cloth from fatty matters, but in modern times other substances have been substituted, and the consumption of it has greatly fallen off.

COAL (MINERAL). Massive, uncrystalline. Color, black or brown; opaque. Brittle or imperfectly sectile. Hardness 0.5 to 2.5. Specific gravity 1.2 to 1.80. Coal is composed of carbon with some oxygen and hydrogen, more or less moisture, and

14

traces also of nitrogen, besides some earthy material which constitutes the ash.

Anthracite (*Glance coal, Stone coal*). Lustre high, not resinous, sometimes submetallic. Color, gray-black. Hardness 2 to 2.5. Specific gravity, if pure, 1.57 to 1.67. Fracture often conchoidal. Good anthracite contains 78 to 88 per cent. of fixed carbon.

Bituminous coal. Color, black. Lustre, usually somewhat resinous. Hardness 1.5 to 2; specific gravity 1.2 to 1.4. Contains usually 75 to 85 per cent. of carbon.

Cannel coal. Very compact and even in texture, with little lustre, and fracture largely conchoidal.

Brown coal (often called *lignite*). Color, black to brownish black. Contains 52 to 65 per cent. of fixed carbon.

Jet resembles cannel coal, but is harder, of a deeper black and higher lustre. It takes a brilliant polish and is set in jewelry.

DOLOMITE is composed of carbonic acid, lime, magnesia. It occurs in rhombohedrons, faces often curved. It is frequently granular or massive; white or dull tinted; and glassy or pearly. Specific gravity 2.8 to 2.9; hardness 3.5 to 4. Effervesces in nitric acid and dissolves more slowly than calc spar. Yields quicklime when burnt. Occurs in extensive beds of various ages like limestone. It is used as a building-stone and in the manufacture of Epsom salts. It is difficult to distinguish from calcite without chemical analysis.

FELDSPAR, ORTHOCLASE, is composed of silica 64.20, alumina 18.40, potash or soda (lime) 16.95. Crystallized or in irregular masses. Opaque; usually flesh red or white, or of various dull tints. Lustre, glassy or pearly ; fracture, irregular, but in some directions it splits with an even, glimmering cleavage face. Specific gravity 2.3 to 2.8 ; hardness 6. Before the blow-pipe it fuses with difficulty ; is not touched by acids. Where found in sufficient quantity to be of industrial value, it is usually obtained from veins in granite or pegmatite. The minerals associated with feldspar are chiefly quartz and mica, while tourmaline and topaz also occur commonly. Feldspar is, to a limited extent, employed in the manufacture of glass, but the chief use for it is as a china glaze and as a glass-forming ingredient in the body of the porcelains.

One of the finest varieties of feldspar is that known as *Adularia*, from Mount Adula near the St. Gothard Pass, where it is found redeposited from the rock mass in veins and cavities. It consists of silica 64, alumina 20, lime 2, and potash 14. *Moonstone* is another variety with bluish-white spots of a pearly lustre. *Sunstone* is another with a pale yellow color with minute scales of mica. *Aventurine*, feldspar sprinkled with iridescent spots from the presence of minute particles of titanium or iron. The last three varieties are employed as gem-stones, being occasionally set in brooches, but are too soft for rings.

A beautiful variety of orthoclase known as *Ama-*

zon stone occurs in large green crystals near Pike's Peak, in Colorado, in Siberia and elsewhere.

FLINT consists of silica, which in a very fine condition has been separated from the surrounding rock, and which, attracted to some organic or inorganic nucleus, and sometimes only to itself, has grown in successive layers or bands, often of different colors. *Hornstone* or *chert* is allied to flint, but it is more brittle and it takes its color—dirty grey, red, and reddish-yellow, green or brown—from the rocks in which it is found. It occurs in portions of sandstone rocks usually containing a little lime, the fine silica being seemingly collected into one spot.

FLUORSPAR, FLUORITE, consists of 48.7 per cent. of fluorine and 51.3 per cent. of calcium. It occurs in cubes or octahedrons, and also in masses. It is transparent or opaque; white or light violet, blue, green or yellow; sometimes layers of different tints in the same piece. Lustre, glassy. It breaks with smooth cleavage planes parallel to the octahedral faces. Specific gravity 3 to 3.2; hardness 4. Before the blow-pipe it is fusible with difficulty to an enamel. It is used in the manufacture of hydrofluoric acid, with which glass is etched, and also as a flux for copper and other ores. Sometimes it is employed for ornaments, especially massive pieces, they taking a high polish. It occurs in veins with lead and silver ores.

GRAPHITE, PLUMBAGO, BLACKLEAD, consists essentially of carbon, in mechanical admixture with varying proportions of silicious matter, as clay, sand or limestone. It occurs in hexagonal crystals, but usually in foliated or massive layers. Color, steel gray to bluish black. Hardness very slight, 0.5 to 1. Soils the fingers, makes a mark upon paper, and feels greasy. The specific gravities of different kinds of graphite vary according to the content of foreign admixtures, but lie within the limits of 2.105 and 2.5857. Graphite is not affected by acids and strongly resists other chemical agents. It is largely used in the manufacture of pencils, crucibles, stove polish, and lubricants for heavy machinery. It is found in various parts of the world, chiefly in crystalline limestone, in gneiss and mica schists, frequently replacing the mica in the latter so that they become actual graphite schists. The chief source whence the bulk of the mineral has for many years been derived is the Island of Ceylon. In the United States graphite is obtained from a mountain, locally known as the Blacklead Mountain, which rises close to the village of Ticonderoga, Essex Co., New York. The graphite beds are interstratified between gneissic rocks. The beds dip at an angle of 45°. The ore in them is chiefly of the foliated variety, and is mixed with gneiss and quartz in the beds in veins or layers from 1 to 8 inches in thickness, some of the deposits being richer than others. One of these has been followed to a depth of 350 feet. It is found of varying thickness and it opens out at times into pockets.

Graphite is said to occur in great purity in different localities in Albany Co., Wyoming, in

veins from 1 foot 6 inches to 5 feet thick. At
Pilkin, Gunnison Co., it occurs massive in beds 2
feet thick, but of impure quality. It is also found
in the coal measures of New Mexico, in Nevada, in
Utah, and in the Black Hills of South Dakota.

The value of graphite depends upon the amount
of its carbon. To test the purity of graphite, pulverize and then dry at about 350° F. 20 grains
of it; then place it in a tube of hard glass 4 to 5
inches long, half an inch wide, and closed on one
end. Add twenty times as much dried oxide
of lead and mix intimately. Weigh the tube and
contents, and afterwards heat before the blow-pipe
until the contents are completely fused and no
longer evolve gases. Ten minutes will suffice for
this. Allow the tube to cool, and weigh it. The
loss in weight is carbonic acid. For every 28 parts
of loss there must have been 12 of carbon.

GYPSUM is a hydrous sulphate of lime, and is
composed of sulphuric acid, lime and water. It
occurs in prisms with oblique terminations, sometimes resembling an arrow-head. It is transparent
or opaque, white or dull tinted, with a glassy,
pearly or satin lustre. Cleavage occurs easily in
one direction; specific gravity 2.3; hardness 2; can
be readily cut with the knife. In the blow-pipe
flame it becomes white and opaque without fusing,
and can then be easily crumbled between the
fingers. Nitric acid does not cause effervescence.
It occurs in fissures and in stratified rocks, often
forming extensive beds. When pure white it is

called ALABASTER; when transparent, SELENITE; and when fibrous, SATIN SPAR. When burnt, gypsum loses its water and falls to powder. This powder, called PLASTER OF PARIS, which is perfectly white when free from iron, possesses the property of reabsorbing the water lost, and in a very short time of assuming again the solid state, expanding slightly in so doing. It is this last property that renders plaster of Paris so valuable for obtaining casts. It is also used as a fertilizer.

INFUSORIAL EARTH is an earthy, sometimes chalk-like siliceous material, entirely or largely made up of the microscopic shells of the minute organisms called *diatoms*. It occurs in beds sometimes of great extent, sometimes beneath peat beds, and is obtained for commerce in Maine, New Hampshire, Massachusetts, Virginia, California, Nevada, Missouri. It feels harsh between the fingers and is of a white or grayish color, but often discolored by various impurities. Infusorial earth is used as a polishing powder, electro-silicon being the trade-name of one kind much used for polishing silver. It is also used for making soda silicate and for purposes of a cement. Being a bad conductor of heat, it is applied as a protection to steam boilers and pipes. It is also employed for filling soap.

LITHOGRAPHIC LIMESTONE. The only stone yet found possessing the necessary qualifications for lithographic work is a fine-grained homogeneous limestone, breaking with an imperfect shell-like or conchoidal fracture, and, as a rule, of a gray, drab or yellowish color. A good stone must be sufficiently porous to absorb the greasy compound which holds the ink, soft enough to work readily under the engraver's tool, yet not too soft, and must be firm in texture throughout and entirely free from all veins and inequalities. The best stone, and indeed the only one which has yet been found to fill satisfactorily all these requirements, occurs at Solenhofen, Bavaria. These beds are of Upper Jurassic age, and form a mass of some eighty feet in thickness. The prevailing tints of the stone are yellowish or drab.

In the United States materials partaking of the nature of lithographic stone have been reported from various localities, but we believe all have failed as a source of supply of the commercial article, though it is possible that ignorance as to the proper methods of quarrying may in some cases have been a cause of failure.

MEERSCHAUM or SEPIOLITE is a manganese silicate. When pure, it is very light; and, when dry, it will float upon water. It will be recognized by its property, when dry, of adhering to the tongue, and by its smooth, compact texture. It is generally found in serpentine, in which rock it occurs in nodular masses; but it is also found in limestones of tertiary age. It is of a snowy-white color and a useful substance when found in quantity, being much employed for the bowls of tobacco pipes, and for this purpose is mined in Asia Minor.

MICAS. These are silicates of alumina with potash, rarely soda or lithia, also magnesia, iron and some other elements. Always crystallized in thin plates, which may be split into extremely thin flexible layers. Transparent in thin layers. Color, white, green, brown to black. Specific gravity 2.7 to 3.1. Hardness 2 to 2.5; very easily scratched with a knife. Before the blow-pipe it whitens, but is infusible except on thin edges. When it can be obtained in large sheets, mica is very valuable. It is sometimes used in the place of window glass on board ship, for stoves and for chimneys for lamps. The ground material is used as a lubricant and in making ornamental and fire-proof paint.

Biotite, or black mica, contains more magnesia than alumina. It is often present in eruptive rocks, especially some granites. *Muscovite,* or white mica, on the contrary, contains more alumina than magnesia, and as it also contains potash in small but appreciable quantities, it is sometimes called *potash mica,* and biotite *magnesian mica.* Muscovite is an important mineral to the tin miner, since it is always found in that class of granite in which tinstone occurs, and with quartz alone forms the rock called *greisen,* which is very generally associated with tin. The rock in which large sheets of mica are found is called by some geologists *pegmatite,* and has the same composition as granite itself, but the crystals are of a larger size.

MOLYBDENUM. The sulphide occurs native as MOLYBDENITE in crystallolaminar masses or tabular crystals, having a strong metallic lustre and lead-gray color, and forming a greenish-black streak which is best seen by drawing a piece across a china plate. Specific gravity 4.5 to 4.6 ; hardness 1 to 1.5 ; easily scratched by the nail. It contains 58.9 of molybdenum and 41.1 per cent. of sulphur. It occurs sparingly in granite, syenite and chlorite schists, and is sometimes mistaken for graphite, from which it is, however, readily distinguished by the streak, that of graphite being black. Before the blow-pipe it is infusible, but tinges the flame faint green. Heated on charcoal for a long time it gives off a faint sulphurous odor and becomes encrusted white. Its chief use is in the preparation of a blue color.

NITRE or *saltpetre* is composed of potash and nitric acid. It is soluble in water. It has a cooling taste, and is easily distinguished by the vivid manner in which it burns on red-hot charcoal. It is usually found native as an efflorescence on the soil.

ROCK SALT has the character of ordinary table salt, but is more or less impure. Occurs in beds interstratified with sandstones and clays, which are usually of a red color and associated with gypsum. Specific gravity, 2 to 2.25 ; hardness, 2 to 2.5. It contains 39.30 per cent. of sodium and 60.66 per cent. of chlorine, but most samples contain clay and a little lime and magnesia. The surface indications of rock salt are brine springs supporting a vegetation like that near the sea coast, also occasional sinking of the soil caused by the removal of the subterranean bed of salt by spring water. Rock salt is obtained by sinking wells, from which the brine is pumped and evaporated in large pans, or by mining, the same as for any other ore.

Salt deposits occur in the strata of all ages, from the Silurian to those now forming. In North America a chain of mountains extends along the west bank of the river Missouri for a length of 80 miles by 45 in breadth, and of considerable height. These mountains consist largely of rock salt. The same formation extends into Kentucky, where the deposits are called "licks," because of the licking of the rocks and soil by the herds of wild cattle that once roamed there. In Michigan, in the neighborhood of Marine City, a well was sunk to a depth of 1,633 feet, when a deposit of rock salt was entered and penetrated to a depth of over 1500 feet without the tools passing through it. The deposit seems to increase in thickness, but it is reached at an increasing depth as it trends in a south-westerly direction by Inverhuron, Kincardine, and War- wick.

An extraordinary superficial deposit of rock salt occurs in Petite Anse Island, parish Iberia, Lou- isiana. The island is about two miles in diameter, and the salt deposit on it is known to extend under 165 acres. It is covered with 16 feet of soil. It has been proved to a depth of 80 feet. The salt occurs in solid masses of pure crystals, and it is taken out by blasting.

The bulk of the manufactured salt in North America is obtained from brine springs. Valuable and productive springs are worked in Syracuse and Salina districts, New York, and in Ohio. Some of these arise from a red sandstone whose geological place is said to be below the coal measures. Rock salt has been discovered in Nevada. The southern termination of the deposits is about seven miles from the uppermost limit to the navigation of the Colorado river. Some of the specimens are sufficiently pure and transparent to allow of small print being read through them. In the same state there is an interesting salt lake, the water of which contains about two pounds of salt and soda to every gallon. It is several hundred feet deep. Soda and salt have been obtained from this lake for several years by natural evaporation. The water is pumped into tanks at the beginning of the summer season. It is left in these tanks during the warm summer months until the frost sets in. When the first frost comes the soda is precipitated in crystals. The water is then drained off into a large pond, where slow evaporation goes on, and a deposit of common salt is obtained.

The famous salt mine of Wieliezka, near Cracow, in Galicia, has been worked since the year 1251, and it has still vast reserves of the mineral.

SLATE is an argillaceous shale easily recognized by its cleavability, and varies in color from light sea-green and gray to red, purple and black. It has been formed by sedimentary deposits, and now constitutes extensive beds in the Silurian formation.

SULPHUR. Native sulphur or brimstone occurs

crystallized or massive in volcanic regions and in beds of gypsum. Color, yellow; lustre, resinous; specific gravity 2.1; hardness 1.5 to 2.5. It is fusible and burns with a blue flame and well-known odor. It is frequently found contaminated with clay or pitch. Italy and Sicily together furnish the greater part of the sulphur of commerce, the major portion coming from Sicily. The most important deposits of brimstone in the United States are found in Utah at Cove Creek, 22 miles from Beaver, while there are other deposits at a point about 12 miles southwest from Frisco. Large deposits of sulphur are known to exist in Wyoming, California and Arizona, but none of them is at present available for working at a profit.

A scarcity of brimstone has led to greater attention being paid to native pyrites, especially for the manufacture of sulphuric acid. While there are many deposits of iron pyrites in most parts of the world, they are not always accessible to mining at a low cost, and situated so that transportation of the low-valued product is easy and cheap. These primary conditions are essential to the industrial usefulness of any pyrites bed. The production of pyrites on a commercial scale in the United States is at present confined to Massachusetts and Virginia.

As a rapid and accurate method of estimating the sulphur available to the acid maker in a sample of pyrites, J. Cuthbert Welch has published the following in the *Analyst*: Place 5 grammes of pyrites in a porcelain boat in a combustion tube, heat to redness, pass oxygen * over till combustion is complete, and absorb the gas formed in about 30 cubic centimeters of a solution of bromine in a mixture of equal parts of hydrochloric acid (specific gravity 1.1) and water, in potash (or preferably nitrogen) bulbs. Wash out the solution into a beaker, boil, precipitate by boiling solution of barium chloride, cool, filter, and wash, dry and ignite the barium sulphate.

TALC or SOAPSTONE, called STEATITE when massive, is a silicate of magnesia. It is trimetric, foliated or massive, nearly opaque, of a white or green color, pearly lustre and greasy feel. Specific gravity 2.7 ; hardness 1 ; easily impressed by the nail, but impure varieties are much harder. It is readily distinguished by its greasy feel and pearly lustre ; it is not attacked by boiling sulphuric acid. It is often applied to useful purposes, as for gas burners, a filling for paper, etc.

---

* The oxygen should be prepared from pure potassium chlorate in glass vessels, or at any rate in an iron one, kept especially for the purpose, and the gas should be passed through a strong solution of potash in the bulbs, through a U-tube containing calcium chloride, and lastly either through another calcium chloride tube or, preferably, over phosphoric anhydride before use.

CHAPTER XIV.

PETROLEUM, OZOCERITE, ASPHALT, PEAT.

CRUDE PETROLEUM occurs only in the higher strata of rocks, it being never found in metamorphic rocks or crystalline formation. The Pennsylvania oil strata belong to the Devonian age, the anticlinal ridges being more favorable, it is said, than the synclinal ones. In Kentucky it occurs near the base of carboniferous limestone. In California it is found in strata belonging to the tertiary age, in Colorado and other western States in those belonging to the cretaceous, and in North Carolina in those belonging to the triassic. In West Virginia it occurs in strata belonging to the coal measures. Crude petroleum is a fluid of a dark color, sometimes black, and contains 84 to 88 per cent. of carbon, the rest hydrogen.

In prospecting for petroleum, the prospector, besides the customary outfit, should carry a stick provided with a long iron point. It is best to follow the courses of rivers and creeks upward, because the progress of the work will not then be impeded by the turbidity of the water. It is also advisable to make such excursions in the warm season of the year, because the oil exudes more freely at that time

(223)

than in cooler weather, when especially heavy oils
and mineral tar, or maltha, are readily converted
into a butyraceous mass.   It is also best to wait
until the water in the rivers and creeks is low.

Observe whether the surface of the water exhibits
variegated iridescent figures, this being especially
the case in places where the water stands quietly or
moves very little, for instance, in coves.   Such an
iridescent film, when found, may be due to petroleum, but also to iron oxides and similar substances.
However, by touching the surface of the water, for
instance, with the iron-pointed stick, a film of oxide
of iron may be disintegrated in angular pieces and
very small flakes, which can be moved in any direction, while oil films, when separated, reunite, and
can be readily distinguished from allied indications
by the many changes in color and figures.   To be
sure, films of very heavy oil may occasionally be
met with which can be separated into angular pieces,
behaving in this respect like iron oxides, but they
almost invariably exhibit variegated movable rings
of color.   In swamps other substances may produce
a phenomenon similar to crude oil.

When indications of oil have in this manner been
discovered in a quiet part of a water-course, try to
remove the iridescent film and turn up the bottom
by several times driving the iron-pointed stick into
it.   If films of oil together with bubbles of gas reappear, and this phenomenon occurs regularly after
repeated experiments, there may be an outcrop of
oil which deserves further examination.

However, if the work with the iron-pointed stick yields negative results, the oil must have floated down from above, and the examination of the water-course has to be continued until by means of the iron-pointed stick the source of the traces of crude oil has been found. This source will usually be in sandstone or other porous rock, and pieces knocked off with a hammer will exhibit the oil generally in the form of drops, partly upon the surfaces of the strata and partly also in small cavities. Instead of petroleum, mineral tar—a black, smeary mass—will frequently be found.'

The rock itself is occasionally impregnated, which may be recognized partly by the odor and partly by the so-called *water-test*. For this purpose place a piece of the rock in quiet water, if possible exposed to the rays of the sun ; if the rock contains oil the characteristic iridescent colors appear, as a rule, immediately upon the surface of the water.

The fresh fracture of oil-bearing sandstone is, as a rule, of a darker color than that of adjoining rock. After rain, drops of water adhere to out-crops of oil sandstone in a manner similar to that observed on other fatty substances.

If in prospecting in water-courses oil-bearing sandstone has been found, the question has to be answered whether the prospector has to deal with contiguous rock or simply with an erratic block. This question can, as a rule, be decided without much difficulty, from the position of the stratification and the petrographic character of the rock in

15

question as compared with the surroundings. However, if there is still a doubt, examine, by means of the water-test, the portions of rock in the natural continuation of the block.

Should the oil-bearing rock actually turn out to be an erratic block, the rock from which it has been derived will be found above, either on the slopes or in the water-course itself. Knowing the petrographic character of the oil-bearing block, it will not be difficult to find in the neighborhood the rock from which it is derived. In the above-described manner the water-courses are traced to the limits of the territory. In carrying on the work of prospecting, it is advisable to examine specimens of all the sandstone by means of the water-test, since the latter frequently shows the presence of petroleum, though there may be no external indications of it.

It may be mentioned, that in cooler weather the traces of oil upon the surface of the water do not yield blue, red, yellow, etc., figures, or at least not very vivid ones, but a milky coloration, which possibly may also be due to other causes, so that determination is more difficult and less certain. This is another reason why it is advisable to select warm days for prospecting. That oil may also be detected by its odor need scarcely be mentioned.

In *swampy puddles* iridescent films, which do not consist of iron oxides, but of hydrocarbons formed by decomposition, are occasionally met with. If due to the latter cause, they do not reappear, or at least only to a slight extent, when removed with the iron-pointed stick from the surface of the water. However, in examining the bottom, gas-bubbles generally rise to the surface. Such puddles are examined first in the centre, and then by detaching pieces from the edges with the iron-pointed stick.

SALSES (*mud-volcanoes*), as well as abundant exhalations of natural gas, if not derived from coal measures, are promising indications of the presence of petroleum in the territory.

It need scarcely be mentioned that porous rock— if oil-bearing—justifies greater expectations than compact rock, and that larger quantities of oil may be looked for in oil-bearing sandstone of greater thickness.

Although, generally speaking, a rich occurrence of oil may be inferred from abundant indications in the outcrop, the reverse is not always correct; in many oil-fields, now productive, the indications when first found were not especially encouraging.

If the oil occurs in definite geological horizons, the latter must be particularly searched for and traced and carefully examined in the water-courses crossing them, not only because the strata are there most denuded so as to allow of the best view of their geological structure, but also because the oil, since the restraining covering is wanting, has the best chance of exuding there, and the cut of the watercourse is generally one of the lowest points of the outcrop, where the most abundant exudation takes place in consequence of the greater head of pressure.

A very important question is whether the oil

occurs in beds or in veins.   In answering this question the following particulars may serve as guiding points.

With proportionately great denudation of the oil-bearing rock, it is sometimes possible directly to decide this question by observation, whereby the prospector, however, must take into consideration that even with a bed-like occurrence the oil will collect in small fissures. With a vein-like occurrence a fissure may be traced to where it assumes larger dimensions in the strike and dip.

If the prospector has to deal with a thick seam or stratum of sandstone, recognized as oil-bearing, imbedded in another rock, for instance, shale, such seam should be traced and pieces freshly cut from it examined as to their content of oil by the water-test. If positive results are obtained, it may be inferred that the sandstone is the bearer of the oil, and that it is a bed-like occurrence.

In a large mass of sandstone several outcrops of oil may sometimes be found at quite a distance from each other.   If in tracing the stratum of the first outcrop according to its strike, the second, third, etc., outcrops are encountered, we have to do with a bed-like occurrence. This tracing of the stratum is effected by means of a compass, however, always with due consideration to the configuration of the ground. Suppose the cross-section of the sandstone bed with the declivity—the so-called out-crop-line—construed and traced.   The outcrop-line will deviate the more from the straight line of strike, the flatter the strata and declivities lie. In tracing the same stratum, it must be observed whether its strike does not change, which, of course, will necessitate a change in the route of the prospector.

If some promising outcrops of oil have been found, which will justify the execution of more extensive and more expensive prospecting work, it is advisable to mark accurately in the sketch-map, in addition to the outcrops, the relative heights, generally determined by an aneroid barometer, the strike and dip of the stratum reduced to the astronomical meridian, and the outcrops of well characterized concordant strata, for instance, imbedded shale, S, Fig. 55, no matter whether they lie in the upcast or downcast of the outcrops of oil, a. The relative heights of one of these strata are determined in several places, selecting points which can be readily found upon the map, and, if possible, lie at the same height, which can be readily effected without essential error with the assistance of an aneroid barometer by taking observations in rapid succession. The points of same height, for instance, 1 and 2, give the strike of the stratum for a greater distance.

By connecting the outcrops of oil a by a line AA, and again determining in the latter several points of the same height, for instance, 3, 4 and 5, the general strike is again obtained. If the latter runs parallel with the general strike of the characteristic stratum S, previously traced, one is justified in inferring a bed-like occurrence of oil, even if the construed dip of the outcrop line of oil corresponds with the observed local dip of the strata.

In these investigations it is presupposed that the oil is recognized as exuding from the solid rock, an error regarding the outcrop of it being, therefore,

FIG. 55.

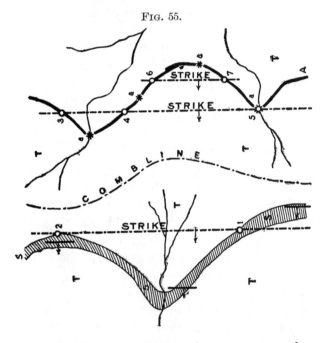

excluded. Such an error may, however, occur when the outcrop is covered with loose masses of earth and rock, to the base of which the oil exuding above flows down hidden, and escapes further below by some accidental cause.

A vein-like occurrence of oil will not show the

above-mentioned conformities with the characteristic concordant strata. Such an occurrence presupposes a fissure, which is generally connected with a throw of the strata. This is most frequently established by the fact that a characteristic stratum suddenly ends and does not reappear in its natural continuation, but either to the right or left, or higher or lower. If two or more such points of disturbance have been found, their connecting line is the outcrop line of the fissure, Fig. 56. If this line passes

FIG. 56.

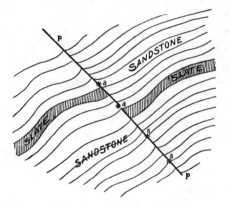

through the outcrop *a*, or if several outcrops lie in it, a vein-like occurrence of oil must be inferred.

However, sometimes the oil occurs in a maze of smaller and larger fissures. This is shown in the construction by the fact that in the presence of several outcrops a linear distribution of the same cannot be recognized, and that the conbinations yield the most varying results, according to whether exploration is carried on from the one or the other outcrop. Such occurrence presents uncommon difficulties in prospecting.

It need scarcely be mentioned that in prospecting for oil, it is of great importance to hunt up and map the anticlinals and their saddles, as well as faults.

The directions here given for prospecting may have to be modified according to local conditions. With a sufficient preliminary knowledge of geology, any difficulties will, as a rule, be readily overcome by thoroughly digesting the principles of the directions given.

As regards the quality of the surface oil, it must be remembered that it is not a criterion for the oil occurring at greater depth. The oil thickens on the surface of the earth, and with increasing density becomes viscous and dark. If pale, limpid, and specifically lighter oil is found at the outcrop, it is sure evidence of oil of excellent quality at greater depth. In every case it may be expected that the quality of the oil at greater depth is superior to that at the outcrop.

OZOCERITE is a mineral paraffine or wax, and occurs generally in fissures and cavities in the neighborhood of coal-fields and deposits of rock salt, or under sandstone pervaded with bitumen. It is found in various localities in Africa, America, Asia and Europe. In the United States it occurs in Arizona, Texas and Utah.

The most interesting deposit is in East Galicia.

The ozocerite occurs there in a saliferous clay belonging to the miocene of the more recent tertiary period, and forming a narrow, almost continuous strip on the northern edge of the Carpathian Mountains. This miocene group of saliferous clay consists chiefly of bluish and variegated clays, sands and sandstones, with numerous occurrences of gypsum, rock salt and salt springs. In Boryslaw, the strata of saliferous clay form a perceptible saddle as they sink on the south below the so-called menilite

FIG. 57.

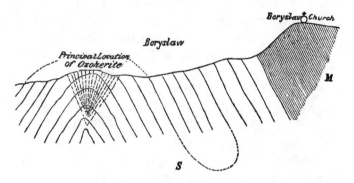

slates, which are very bituminous and foliated, and form here the most northern edge of the Carpathian Mountains. The principal deposit of ozocerite converges with the axis of this saddle as shown in Fig. 57, *S* being the strata of saliferous clay; and *M* menilite slate.

Closely allied to ozocerite are the following mineral resins:

RETINITE, generally of a yellowish-brown, sometimes of a green-yellow or red color. It is found
with brown coal in various localities.

ELATERITE or elastic bitumen, of a blackish
brown color, subtranslucent, and occurring in soft,
flexible masses in the lead-veins of Castleton, in
Derbyshire, in the bituminous sandstone of Wood-
bury, Connecticut, etc.

PYROPISSITE occurs in strata in brown coal.

Ozocerite occurs in various shades of color, from
pale yellow to black; when melted it generally
shows a dark-green color. The pale varieties are
chiefly found in places containing much marsh gas.
The dark-green, heavy variety is the best, while the
black kind, or asphaltic wax, is the poorest; it contains resinous combinations of oxygen, and is intermediate between mineral oil and ozocerite.

The odor of ozocerite is, according to its purity,
agreeably wax-like. In consistency it is soft,
pliable, flexible to hard; the mass in the latter case
showing a conchoidal fracture, but softens on
kneading. The boiling-point lies between 133°
and 165° F., and of the so-called "marble wax"
even as high as 230° F. The specific gravity is
from 0.845 to 0.930.

Ozocerite is readily soluble in oil of turpentine,
petroleum, benzine, etc., and with difficulty in
alcohol and ether; it burns with a bright flame,
generally leaving no residue. Its elementary composition is about that of petroleum, 85 per cent. of
carbon and 15 per cent. of hydrogen.

NATIVE ASPHALT or BITUMEN is solid at the ordinary temperature, of a black to blackish-brown color and a conchoidal fracture with glossy lustre. Hardness 1 to 2 ; specific gravity 1 to 2. It melts at 90° F., and is very inflammable. It appears to be formed by the oxidation of the non-saturated hydrocarbides in petroleum. The most remarkable deposits are in Cuba, Trinidad, and Venezuela. Other noted localities are the Dead Sea, Seyssel (France), Limmer, the Abruzzo, and Val de Travers. It occurs also of every degree of consistence, and in immense quantity, along the coast of the Gulf of Mexico, chiefly in the States of Tamaulipas, Vera Cruz and Tabasco, where not unfrequently it is associated with rock salt and "saltpetre." It also occurs in Utah in widely separated places. It has been found associated with ozocerite and more extensively as melted out of sandstone. California includes a large area which furnishes asphalt, much the larger proportion being the product of the decomposition of petroleum, while the remainder occurs in veins that are evidently eruptive, the former occurring in beds of greater or less extent on hill-sides or gulch slopes, below springs of more fluid bitumen. These deposits are scattered over the country between the bay of Monterey and San Diego, but are chiefly observed west and south of the coast ranges, between Santa Barbara and the Soledad pass. Asphalt occurs also in other localities in the United States, for instance in Connecticut, in thin seams and veins in eruptive rock ; in New

York in the region of eruptive and metamorphic rocks, in Tennessee in the Trenton limestone, etc. In some American specimens sulphur has been found to the extent of 10.85 per cent. Asphalt is in great request for paving purposes; it is of increasing value, and deposits are eagerly sought for.

PEAT. Peat is not a mineral, but consists of the cumulatively resolved fibrous parts of certain mosses and graminaceæ. It gradually darkens from brown to black with increasing age. It occurs in beds or in bogs. As a fuel, it is most economically used at the place where it is grown. Good peat yields about 3 to 6 per cent. of tar proper, which is comparatively easy to purify by the usual method.

CHAPTER XV.

ALTHOUGH many varieties of gems and precious stones are known to occur in the United States, systematic mining for them is carried on only at a few places, and the annual output is still very small in comparison with the prospective extent of the field. Not many persons are familiar with the appearance of gem stones in their native state, so that while quartz pebbles are often mistaken for rough diamonds, garnets for rubies, ilmenite for black diamonds, etc., on the other hand it is quite probable that many valuable occurrences have escaped notice.

DIAMOND. Diamonds are usually met with in alluvial soil, often on gold-diggings. In some Indian fields a diamond-bearing conglomerate occurs which is made up of rounded stones cemented together, and lies under two layers, the top one consisting of gravel, sand and loam, the bottom one of thick clay and mud. In the neighborhood of Pannah, between Sonar and the Sona river, diamonds are found in ferriferous pebble conglomerate and in river alluvium. The most beautiful crystallized specimens are, however, found on the west side of the Nalla-Malla mountains near Ban-

(237)

ganpally, between Pennar and Kistnah, in a diamond-bearing layer between beds of primitive conglomerate.

In Borneo, the diamond is found associated with magnetic iron ore, gold and platinum, in alluvial deposits consisting of serpentine and quartz fragments as well as marl.

In Brazil, the province Minas Geraës is rich in diamonds, the most important occurrence being at Sao Joao do Barro, where they are found in an entirely weathered talcose slate. In other parts of the same country the diamond is also obtained from a conglomerate of white quartz, pebbles and light colored sand, sometimes with yellow and blue quartz and iron sand. In the province of Bahia occurs a substance known as *carbonado* or *black diamond*. It is an allotropic form of carbon closely related to the diamond, and is found in small irregular crypto-crystalline masses of a dark gray or black color. Although its density is not so great as that of the diamond, it is very much harder; in fact, it is the hardest substance known. At first it was used only in cutting diamonds, but since the invention of the core-drill for boring in rocks it has found a greatly extended use, and is now employed for the so-called " diamond crown " of this drill. The *bort* of the South African mines finds a similar industrial application, being worthless as a gem.

In South Africa the diamond occurs associated chiefly with garnet and titanic iron ore, as well as with quartz opal, calcareous spar, and more rarely with iron pyrites, bronzite, smaragdite and vaalite. According to St. Meunier the South African diamond-bearing sands are composed of an exceedingly large number of constituents, eighty different varieties of minerals and rocks having been found in them. Of minerals occur, for instance, diamond, topaz, garnet, bronzite, ilmenite, quartz, tremolite, asbestus, wallastonite, vaalite, zeolite, iron pyrites, brown iron ore, calcareous spar, opal, hyalite, jasper, agate, clay. Of rocks are found, serpentine, eklogite, pegmatite and talcose slate. At the Kimberley mine, which more or less represents others in the neighborhood, the diamond-bearing ground forms a "pipe" or "chimney" surrounded by formations totally different from the payable rock. The encasing material is made up of red sandy soil on the surface, underneath which is a layer of calcareous tufa, then yellow shale, then black shale, and below this, hard igneous rock. The diamond-bearing ground consists of "yellow ground" (really the decomposed "blue ground"), which is comparatively friable; and deeper down the "blue ground" (hydrous magnesian conglomerate), which needs blasting by dynamite. The "blue ground" is of a dark bluish to a greenish gray color, and has a more or less greasy feel. With it are mixed portions of boulders of various kinds of rocks such as serpentine, quartzite, mica-schist, chlorite-schist, gneiss, granite, etc. All this "blue ground" has evidently been subjected to heat. The gems are in the matter which binds these rocks, not in the rocks themselves.

Diamonds are also found in the Ural, various parts of Australia, New Zealand and in the United States. In the latter country diamonds have been found at a number of localities, but never enough to warrant any extended mining for them. Many experienced geologists hold to the opinion that since so many associations of the diamond are present in North Carolina they have hopes of their being found there. The garnet districts of Arizona and New Mexico may also be looked upon as favorable for the occurrence of this gem. Of the localities where diamonds have been found in the United States may be mentioned : The gold diggings of Twitty's mine in the itacolumite region of Rutherford Co., North Carolina, 1847 ; further in Hall Co., Georgia, 1850, in the gold diggings on the south slopes of the Alleghany mountains, in Arizona, and in California, together with platinum in various gold diggings. Further at Dysartville, McDowell Co., North Carolina, in Idaho, San Juan Co., Colorado, and Cherokee Flat and several other localities in Butte Co., California.

The natural surface of the diamond is often unequal ; its sides are lined, somewhat convex, and generally appear dulled, or as they are commonly called, *rough*, by the evident action of fire. The diamond breaks regularly into four principal cleavages. It does not sparkle in the rough, and the best test is its hardness and its becoming electric, when rubbed before polishing. The color of the diamond varies through all tones of the color-scale, from absolute colorless through all shades of yellow, red, green, blue to intense black. Some colorless diamonds acquire on heating a reddish shade, which disappears on cooling.

The occurrence of diamonds of different colors affords a remarkable illustration of what has been said about the colors of minerals. As pure carbon, diamond is colorless, as are also the microscopic diamonds artificially produced by an electric current, but in nature the stones are of different colors, which are imparted to them by a very small proportion of foreign matter. The yellow and gray tints decrease the value of the diamond, but red, blue and green varieties, on the contrary, are so rare, that when diamonds are so colored their value is considerably greater than if perfectly colorless. For instance, the best blue diamond known is estimated at double the calculated value of a good colorless diamond of the same size.

In Borneo a kind of black diamond is found which is very highly prized in consequence of its exceptional lustre and rarity. It is even harder than the ordinary diamond.

The specific gravity of the pure diamond varies from 3.5 to 3.6 ; that of the black diamond is from 3.012 to 3.255.

One of the most beautiful qualities of the diamond is its power of reflection ; that of water is, 0.785 ; that of the ruby, 0.739 ; that of the rock crystal, 0.654 ; that of the diamond, 1.396. The refraction of the diamond is single in the entire crystals ; when

16

broken it possesses double, but imperfect refraction, in the thin layers.

The value of the diamond is dependent on its

FIG. 58.

color, its size and the finish given to it by working. Perfectly colorless stones bring the highest price, and next stones with a reddish, greenish and bluish shade, which, however, are quite rare. Yellowish diamonds are of less value, the price paid for them being the lower the more the yellow color plays into brown.

Of the largest diamonds each has its own name and its own history. Of these may here be mentioned the *Koh-i-noor* or mountain of light, Fig. 58, *d*. It weighs $106\frac{1}{16}$ carats. The *Orlof*, Fig. 58, *a*, weighs $194\frac{3}{4}$ carats, and is as large as half a pigeon's egg ; it adorns the sceptre of the Russian emperor. The *Grand Duke of Tuscany* or *Florentine*, Fig. 58, *b*, is one of the most beautiful diamonds. It is a yellow diamond and weighs $139\frac{4}{8}$ carats. It belongs to the House of Austria. The *Pitt* or *Regent*, Fig. 58, *c*, belongs to the French Treasury and, with the exception of the Koh-i-noor, is the most beautiful and most regular diamond. It weighs $136\frac{3}{4}$ carats.

SAPPHIRE. The sapphire is the blue variety of corundum in its purest crystalline state. Its general composition is alumina 92, silica 5.25, oxide of iron 1.0. The color most highly valued is a highly transparent bright Prussian blue. More frequently the color is a pale blue, passing by paler shades into perfectly colorless varieties. The paler varieties are frequently marked by dark blue spots and streaks which detract from their value. But these paler varieties lose their color when subjected to great heat, a fact that has sometimes been taken advantage of by unscrupulous dealers to pass them off as diamonds.

The principal form of the sapphire is an acute

rhomboid, but it has many modifications and varieties. On being broken it shows a conchoidal fracture, seldom a lamellar appearance. The principal locality for sapphires in the United States is in the garnet districts near Helena, Montana; Santa Fé, New Mexico; southern Colorado and Arizona. Here they occur in the sand, associated with peridot, pyrope and almandine garnet.

RUBY. The ruby is the red variety of corundum and in composition varies from almost pure alumina to a compound containing 10 to 20 per cent. of magnesia, and always about 1 per cent. of oxide of iron. The ruby is subdivided into several varieties according to color, which in its turn is affected by mineral composition, *spinel ruby* occurring in bright red or scarlet crystals, *rubicelle* of an orange red color, *bala ruby* rose red, *almandine ruby* violet, *chlorospinel* green, and *pleonast* is the name given to dark varieties.

The crystals are usually small and when not defaced by friction they have a brilliant lustre, as has also the lamellar structure, with natural joints which it shows on being broken. It exhibits various degrees of transparency. The color most valued is the intense blood red or carmine color of the spinel ruby. When the color is a lilac blue, the specimen was formerly known as the *Oriental amethyst*, and was regarded as a connecting link between the ruby and the sapphire. In the United States the ruby is found in various localities, in some of which the crystals have partly decomposed and show a soft structure resembling steatite. It occurs in gneissic and metamorphic rocks and in granular limestone. In Ceylon it is found with the sapphire in the river deposits.

Topaz is composed of silica, alumina and fluorine. It occurs in prismatic crystals, sometimes furrowed lengthwise, variously terminated, breaking easily across with smooth brilliant cleavage. Transparent or semi-transparent. White, yellow, greenish, bluish, pink. Lustre, glassy. Specific gravity, 3.5. Hardness, 8. Scratches quartz: is scratched by sapphire. Infusible, but often blistered and altered by heat. When smooth surfaces are rubbed on cloth they become strongly electric, and can attract small pieces of paper, but rough surfaces do not show this. The brilliant cleavage of topaz distinguishes it from tourmaline and other minerals. Topaz occurs in gneiss or granite with tourmaline, mica, beryl; also cassiterite or tin-stone, apatite, fluorite. The white topaz resembles the diamond, but unlike the latter it can be scratched by sapphire. The pale blue variety is of value for cutting into large stones for brooches; specimens are occasionally found of several pounds weight. Topaz of a beautiful sherry color occurs in Brazil. Specimens of this when heated become pink, when they are known as burnt topaz. The yellow varieties are cut as gems. Although not very valuable, they are quite brilliant and look very well.

Topaz has been found in Arizona, New Mexico, and occasionally in southern Colorado. In the latter state, and in Utah and Mexico, it sometimes occurs in fine, clear crystals in volcanic rocks. A notable locality, especially for very large crystals, is at Stoneham, Maine, and another at Trumbull, Connecticut.

BERYL or EMERALD is composed of silica, alumina and beryllium or glucinum. It is almost always found in distinct crystals, and usually in forms easy to recognize. The crystals are hexagonal prisms, usually green, transparent or opaque. Lustre, glassy; fracture uneven; specific gravity, 2.7; hardness, 7 to 8; scratches quartz. Infusible, or nearly so, but becomes clouded by heating. Occurs in granite rocks with feldspar and quartz. Valuable for jewelry when transparent and rich grass-green (emerald), or sea-green (aquamarine). Emerald has been found in North Carolina and aquamarine at a number of localities in the United States.

A productive emerald mine is that of Muso, in New Grenada, Mexico. The emerald occurs in veins and cavities in a black limestone that contains fossil ammonites. The limestone also contains within itself minute emeralds and an appreciable quantity of glucina. When first obtained the emeralds from this mine are soft and fragile; the largest and finest emeralds could be reduced to powder by squeezing and rubbing them with the hand. After exposure to the air for a little time they become hard and fit for the jeweler's use.

PHENACITE is a silicate of beryllium or glucinum. Its hardness is about the same as topaz and its specific gravity 3.4 to 3.6. It occurs in glassy rhombohedral crystals, and its hardness, beautiful transparency and color make it valuable for cutting as a gem, since it is capable of extreme polish. Phenacite has been found at Pike's Peak, Colorado, in crystals of sufficient size and quality to furnish fair gems.

ZIRCON is composed of silica and zirconia. It is found in square prisms terminated by pyramids, and in octahedrons, but often also in pebbles and grains. Transparent or opaque. Wine or brownish red, gray, yellow, white. Lustre, glassy; fracture, usually irregular, but in one direction it can be split so as to exhibit a smooth even cleavage face having an adamantine lustre like the diamond. Specific gravity 4.0 to 5.0; hardness 7.5; scratches quartz, is scratched by topaz. Infusible: the red varieties, when heated before the blowpipe, emit a phosphorescent light, and become permanently colorless. Zircon occurs in syenite, granite, basalt. In some regions it occurs in the rock so abundantly that when the rock has been worn down by the weather, it is left unaltered in considerable quantities. It may then be obtained by washing the gravel in the manner of the gold miner. Clear crystals are used in jewelry, in jeweling watches, and imitation of diamond. It may be distinguished from the latter by its inferior hardness, and in not becoming so readily electric by friction. Fine crystals are obtained in New York and Canada; and good specimens also come from North Carolina and Colorado.

GARNET is composed of silica, alumina, lime, iron, magnesia, manganese. It is found almost always in distinct crystals, and as these crystals are commonly isolated and scattered through the rock, it is not difficult to recognize them. The crystals are usually twelve-sided, having the form of a rhombic dodecahedron. They are transparent or opaque; generally red; also brown, green, yellow, black, white. Lustre, glassy or resinous; fracture conchoidal or uneven; specific gravity 3.5 to 4.3; hardness, 6.5 to 7.5; cannot be scratched with a knife. Fusible with more or less difficulty. Red varieties impart a green color to borax bead owing to presence of chromium. Garnet usually occurs in crystals scattered through granite, gneiss or mica schist, also in crystalline limestone; with serpentine or chromite; also in some volcanic rocks. Fine colored transparent varieties (carbuncle, cinnamon stone, almandine) are used in jewelry. The garnets found in New Mexico and Southern Colorado, and there called "rubies," are as fine as those from any other locality, the blood-red being the most desirable. Very fine crystals of cinnamon stone, cinnamon garnet or essonite have been found in New Hampshire, Maine, and at many other points in the United States.

TOURMALINE is composed of silica, alumina, magnesia, boracic acid, fluorine, oxides of iron (lime and alkalies). It is found in prisms with three, six, nine or more sides, furrowed lengthwise, terminating in low pyramids. Commonly black and opaque, rarely transparent, and of a rich red, yellow, or green color. Lustre glassy; fracture uneven; specific gravity 3.1; hardness 7 to 8; cannot be scratched with a knife. When the smooth side of a prism is rubbed on cloth it becomes electric and can attract a small piece of paper. Tourmaline occurs in granite and slate. Only the fine colored transparent varieties, which are used as gems and for optical purposes, are of value. The principal source of tourmaline in the United States is the locality Mount Mica, at Paris, Maine.

EPIDOTE is a silicate of alumina, iron and lime, but varies rather widely in composition, especially as regards the relative amounts of alumina and iron. It is usually found in prismatic crystals, often very slender and terminated at one end only; they belong to the monoclinic system. Lustre, vitreous; color, commonly green, although there are black and pink varieties. Epidote is found in many localities in the United States and in very large crystals ranging from brown to green in color, but as a rule the crystals are only translucent or semi-opaque, though some stones of considerable value and great beauty have been found in Rabun county, Georgia.

OPAL is composed of silica and water. It is never found in crystals but only in massive and amorphous form. Fracture, conchoidal; specific gravity 2.2; hardness, 6; can be scratched by quartz and thus distinguished from it. It is infusible and generally milk-white. The most beautiful variety of opal is that called *precious opal*, which exhibits a beautiful play of colors and is a valuable gem. One kind of precious opal with a bright red flash of light is called the *fire opal*, and another kind is the *harlequin opal*. Common opal does not exhibit this play of colors, and it varies widely in color and appearance. *Milk opal*, as one variety is called, has a pure white color and milky opalescence, while *resin opal* or *wax opal* has a waxy lustre and yellow color. *Jasper opal* is intermediate between jasper and opal; *wood opal* is petrified wood in which the mineral material is opal instead of quartz. Opal is commonly met with in seams of certain volcanic rocks; sometimes it occurs in limestone and also in metallic veins. Precious opal is rare in the United States, though some of high value is said to have been found in Creek Co., near John Davy's River, Oregon.

TURQUOIS is a hydrated phosphate of aluminium, containing also a little copper phosphate which is probably the source of the color, which in the most precious variety is robin's-egg blue, and bluish-green in less highly prized varieties. It occurs only in compact massive forms, filling seams and cavities in a volcanic rock. Specific gravity 3.127. Turquois has been found in the Holy Cross mining region, thirty miles from Leadville, Colorado, and of late years a number of mines have been opened in New Mexico, at Los Carillos and in Grant County. The latter mines produce stones having a faint greenish tinge, which is either due to a partial

change or metamorphism, which has taken place
while the turquois was in the rock, or it may be a
local peculiarity.   Turquois occurs also in Arizona
and at a point in Southern Nevada.   At the latter
place it is found in veins of small grains in a hard
shaly sandstone.   The color of this turquois is a
rich blue, almost equal to the finest Persian, and
the grains are so small that the sandstone is cut
with the turquois in it, making a rich mottled stone
for jewelry.

AGATE is found in almost every part of the world,
and the difference of the constituent parts makes
the specific gravity vary from 2.58 to 2.69.   The
agate, properly so called, is naturally translucent,
less transparent than crystalline quartz, but yet
less opaque than jasper.   It is too hard to be even
scratched by rock crystal.   It takes a very good
polish.   It is never found in regular forms, but
always either in nodules, in stalactites, or in irregular masses.   *Eye agates* consist of those parts of the
stone in which the cutting discovers circular bands
of very small diameter arranged with regularity
round one circular spot.   These circles are frequently so perfect that they appear to be traced by
the compass.   The first round is white, the second,
black, green, red, blue or yellow; the most rare are
those whose circles are at equal distance from the
centre.   *Moss agate* contains brown-black, moss-like
or dendritic forms distributed rather thickly through
the mass.   These forms consist of some metallic
oxide (as of manganese).   Of all the American

stones used in jewelry there is no other of which so much is sold as the moss agate. The principal sources of supply are Utah, Colorado, Montana and Wyoming.

CHALCEDONY is a semi-transparent variety of quartz, of a waxy lustre and varying in color from white through grey, green and yellow to brown. It is translucent or semi-transparent. It occurs in stalactite, reniform or botryoidal masses, which have been formed in cavities in greenstones and others of the older rocks. Into these cavities, as into miniature caverns, water holding silicious matter has penetrated and deposited its solid contents, consisting almost exclusively of silica tinged by the presence of other minerals. Some of these cavities are several feet in diameter, and besides the coloring of the encircling mass there are often, in the interior of the concretions in them, cavities or central nuclea which contain sometimes as many as twenty-four different substances, as silver, iron pyrites, rutite, magnetite, tremolite, mica, tourmaline, topaz, with water, naphtha, and atmospheric air.

CHRYSOPRASE is of a beautiful apple-green color, due to oxide of nickel. In a warm, dry place the color of chrysoprase is destroyed, but it can be again restored by keeping it damp.

CARNELIAN and SARD have red or brownish tints and are varieties of chalcedony.

JASPER is quartz rendered opaque by clay, iron and other impurities. It is of a red, yellow or green color. Sometimes the colors are arranged in ribands, or in other fantastic forms. It is used for ornamental work.

BLOODSTONE or HELIOTROPE is green jasper, with splashes of red resembling blood spots.

ROCK CRYSTAL is pure, transparent, colorless quartz, and is found at a great many localities in the United States. In Herkimer County, at Lake George, and throughout the adjacent regions in New York state, the calciferous sandstone contains single crystals, and at times cavities are found filled with doubly terminated crystals, often of remarkable perfection and brilliancy. These are collected, cut, and, often uncut, are mounted in jewelry and sold under the name of " Lake George diamonds."

AMETHYST is a transparent variety of quartz of a rich violet or purple color due to the oxide of manganese which it contains. It crystallizes in the form of a hexagon, terminated at the two heads by a species of cone with six facets. These crystals are often in masses, and the base is always less colored than the top. Amethysts are generally found in metalliferous mountains, and are always in combination with quartz and agate. They occur in many localities in the United States, but not in as fine or large specimens as in Brazil or Siberia.

ROSE QUARTZ is pink, red and inclining to violet-blue in color. Occurs in fractured masses and is imperfectly transparent. The color is most permanent in moisture.

SMOKY QUARTZ are quartz crystals tinted with a

a smoky color, becoming sometimes black and opaque.

YELLOW OR CITRON QUARTZ or FALSE TOPAZ occurs in light-yellow translucent crystals. It is often set and sold for topaz, but it may be distinguished from it by the absence of cleavage.

ONYX AND SARDONYX. A variety of quartz having a regular alternation of strata more or less even, and variously colored in black, white, brown, gray, yellow and red. When an onyx has one or two strata of red carnelian, it is more valued and takes the name of sardonyx. In the onyx the dark strata are always opaque and contrast strongly with the clear, which, when thinned, become almost translucent.

CAT'S EYE consists of a quartz mixed with parallel fibres of asbestus and amianthus. It is found in pebbles and in pieces more or less round ; it has a concave fracture ; is translucent and also transparent at the edges. It has a vitreous and resinous light. It is generally either green, red, yellow or gray. It marks glass. Its specific gravity is from 2.56 to 2.73. When exposed to a great heat it loses lustre and transparency, but does not melt under the blowpipe unless reduced to minute fragments.

Many other gem stones are known to occur in the United States, and the following list compiled by Mr. George F. Kunz * is here given :

* Mineral Resources of the United States, Washington, 1883.

List of gem stones known to occur in the United States.

Achroïte (tourmaline).
Agate (quartz).
Agatized wood (quartz).
Almandine (garnet).
Amazon stone (microlene).
Amber.
Amethyst (quartz).
Aquamarine (beryl).
Asteria.
Beryl.
Bloodstone.
Bowenite (serpentine).
Cairngorm (quartz).
Catlinite.
Chalcedony (quartz).
Chiastolite.
Chlorastrolite.
Chondroite.
Chrysolite.
Danburite.
Diamond.
Diopside (pyrozene).
Elæolite (nephelite).
Emerald (beryl).
Epidote.
Essonite (garnet).
Flèche d'amour (quartz).
Fluorite.
Fossil coral.
Garnet.
Grossularite garnet.
Heliotrope.
Hematite.
Hiddenite (spodumene).
Hornblende in quartz.
Idocrase.
Indicolite (tourmaline).
Iolite.
Isopyre.

Jade.
Jasper.
Jet (mineral coal).
Labradorite.
Labrador spar (labradorite).
Lake George diamonds (quartz).
Lithia emeralds (spodumene).
Macle.
Malachite.
Moonstone (feldspar group).
Moss agate (quartz).
Novaculite (quartz).
Obsidian.
Olivine (chrysolite).
Opalized wood (opal).
Peridot (chrysolite.)
Phenakite.
Prehnite.
Pyrope (garnet).
Quartz.
Rhodonite.
Rock crystal (quartz).
Rose quartz (quartz).
Ruby (corundum).
Rubellite (tourmaline).
Rutile.
Rutile in quartz (quartz).
Sagenite (quartz).
Sapphire (corundum).
Silicified wood (quartz).
Smoky quartz (quartz).
Smoky topaz (quartz).
Spinel.
Spodumene.
Sunstone (feldspar).
Thetis hair stone (quartz).
Thomsonite.
Tourmaline.

Topaz.
Turquois.
Venus hair stone quartz.
Willemite.
Williamsite (serpentine).

Wood agate (quartz).
Wood jasper (quartz).
Wood opal (opal)
Zircon.
Zonochlorite (prehnite).

List of species and varieties found in the United States, but not met with in gem form.

Andalusite.
Axinite.
Cassiterite.
Chrysoberyl.
Cyanite.

Ilvaite.
Opal.
Prase (quartz).
Sphene.
Titanite.

List of species and varieties not yet identified in any form in the United States.

Alexandrite.
Cat's-eye chrysoberyl.
Cat's-eye quartz.
Chrysoberyl cat's eye.
Chrysoprase.

Demantoid.
Euclase.
Lapislazulite.
Ouvarite.
Quartz cat's eye.

List of gem stones occurring only in the United States.

Bowenite.
Chlorastrolite.
Chondrodite.
Hiddenite.
Lithia emerald.
Novaculite.

Rutile.
Thetis hair stone.
Thompsonite.
Willemite.
Williamsite.
Zonochlorite.

Table of Characteristics of Gems.

Name and Color.	Lustre.	Specific Gravity.	No. in scale of hardness.	Hardness.	Composition.	System of Crystallization.	Form of Crystal.
DIAMOND, white, pink, yellow, red, blue, green, black, orange, brown, opalescent.	Adamantine; reflects prismatic colors.	3.4 to 3.6	10	Scratches all other precious stones.	Pure carbon.	Cubical.	Cube, octahedron, rhombic dodecahedron, tetrahedron, hexa-octahedron.
SAPPHIRE, white, blue, violet.	Vitreous, very lively.	3.9 to 4.2	9	Scratched by a diamond, scratches all others.	Alumina, 98.5; Oxide of iron, 1.0; Lime, 0.5	Hexagonal.	Hexagonal prism; often pointed at each end.
RUBY, pink, red, violet-red. TOPAZ, ORIENTAL, yellow. AMETHYST, ORIENTAL, purple, violet. EMERALD, ORIENTAL, green, generally pale. CHRYSOBERYL or ORIENTAL CHRYSOLITE, bright pale green, greenish-yellow, reddish-brown.	Vitreous, sometimes pearly.	3.0 to 3.8	8.5	Scratched by sapphire, etc.; scratches quartz readily.	Alumina, 80.2; Glucina, 19.8; (Trace of peroxide of iron, of oxide of lead, and copper, depending on color and locality.)	Rhombic.	In flat hexagonal crystals; generally in rolled pebbles.
CYMOPHANE or CHRYSOBERYL, *Cat's Eye*, when showing an opalescence like a cat's eye. SPINEL, dark red, white, blue, green.	Vitreous.	2.8		Scratched by sapphire, scratches quartz readily.	Alumina, 69.01; Magnesia, 26.21; Protoxide of iron, 0.71; Silica, 2.02; Oxide of chromium, 1.10	Cubical.	Octahedron, rhombic dodecahedral octahedron, tri-octahedron.

17

Table of Characteristics of Gems.—Continued.

Name and Color.	Lustre.	Specific Gravity.	Hardness.	No. in scale of hardness.	Composition.	System of Crystallization.	Form of Crystal.
TOPAZ, white, greenish, yellow, orange, cinnamon, bluish, pink.	Vitreous.	3.5 to 3.6	Scratched by sapphire, scratches quartz readily.	8	Silica, 34.01; Alumina, 58.38; Fluorine, 15.06; Traces of metallic oxides.	Rhombic.	Right rhombic prism, octahedral rhombic prism.
EMERALD, fine green.	Vitreous.	2.67 to 2.75	Scratched by spinel, scratches quartz.	7.5 to 8	Silica, 68.50; Alumina, 15.75; Glucina, 12.50; Oxide of iron, 1.00; Lime, 0.25.	Hexagonal.	Hexagonal prism.
BERYL, or AQUAMARINE, pale sea-green, blue, white, yellow, rarely pink.							
HYACINTH or ZIRCON, brownish-yellow, brownish-red, cinnamon.	Vitreous (almost adamantine).	4.07 to 4.70	Scratches quartz slightly.	7.5	Silica, 33.0; Zircoma, 66.8; Peroxide of iron, 0.10.	Tetragonal.	Long and short square prisms. Long square octahedron. The prisms often doubly terminated with square pyramids.
JARGON, various shades of green, yellow, white, brown. GARNET.	Vitreous, inclining to resinous.	3.5 to 4.3	Scratches quartz slightly.	6.5 to 7.5	Silica, 38.25; Alumina, 19.35; Red oxide of iron, 7.33; Lime, 31.75; Magnesia, 2.40; Protoxide of manganese, 0.50.	Cubical.	Rhombic dodecahedron, rhombic dodecahedral cube, trapezohedron, hexa-octahedron.

Table of Characteristics of Gems.—Concluded.

Name and Color.	Lustre.	Specific Gravity.	Hardness.	No. in scale of hardness.	Composition.	System of Crystallization.	Form of Crystal.
ALMANDINE, violet-red. CARBUNCLE, red, brownish. CINNAMON STONE, white, yellow-orange. PYROPE, vermilion or Bohemian garnet.	Rarely asteroid.						
TOURMALINE, green, red, brown, yellow, blue, black, sometimes white	Vitreous.	2.9 to 3.3	Scratches quartz slightly.	7 to 7.5	Fluorine, 2.28; Silica, 38.25; Boracic acid, 8.25; Alumina, 31.32; Red oxide of iron, 1.27; Magnesia, 13.89; Lime, 1.60; Soda, 1.28; Potash, 0.26	Hexagonal.	Obtuse rhombohedron, hexagonal prisms.
TURQUOISE, blue, green, white.	Vitreous.	2.62 to 3	Scratches glass feebly,	6	Phosphoric acid, 27.34; Alumina, 47.45; Oxide of copper, 2.05; Oxide of iron, 1.10; Oxide of manganese, 0.50; Phosphate of lime, 3.41; Water, 18.18	None.	None.
OPAL, colorless, red, white, green, gray, black, yellow, (iridescent).	Vitreous, inclining to resinous.	2.0 to 2.3	Scratches glass slightly.	5.5 to 6.5	Silica, 91.32; Water, 8.68; Traces of mineral coloring matter.	None.	None.

APPENDIX.

WEIGHTS AND MEASURES.

BRITISH weights and measures, and those used in our country, are based upon the weight of a cubic inch of distilled water at 62° Fah., and 30 inches height of the barometer, the maximum density. This was decided by Parliament, in the reign of George IV., to be 252.458 grains. Recent experiments, however, show that a cubic inch of water at the temperature of maximum densisty is 252.286 standard grains. On this account scientists are urging the readjustment of the gallon, bushel, etc., but at present the tables below are correct. *See also No.* 8.

Weights and measures of various nations :—

No. 1.—ENGLISH LENGTH.

3 barleycorns	=	1 inch.
12 inches	=	1 foot.
3 feet	=	1 yard.
5½ yards	=	1 rod, pole, or perch (16½ feet).
4 poles or 100 links	=	1 chain (22 yards or 66 feet).
10 chains	=	1 furlong (220 yards or 660 feet).
8 furlongs	=	1 mile (1760 yards or 5280 feet).

A span = 9 inches; a fathom = 6 feet; a league = 3 miles; a geographical mile = 6082.66 feet, same as nautical knot, 60 being a degree, *i. e.*, 69.121 miles.

Particular Measures of Length.

A point, $\frac{1}{72}$ of an inch.   A pace, military, 2 feet, 6 inches.
A line, $\frac{1}{12}$ of an inch.   A pace, geometrical, 5 feet.
A palm, 3 inches.   A cable's length, 120 fathoms.
A hand, 4 inches.   A degree (average) $69\frac{1}{8}$ miles.
A link, 7.92 inches.

No. 2.—Surface Measure.

144 square inches	=	1 square foot.
9 square feet	=	1 square yard.
30¼ square yards	=	1 pole, rod, or perch (square).
16 poles (square)	=	1 chain (sq.) or 484 sq. yds.
40 poles	=	1 rood (sq.) or 1210 sq. yds.
10 chains or 4 roods	=	1 acre (4840 sq. yds.).
640 acres	=	1 sq. mile.

No. 3.—Surface Measure in Feet.

9 square feet	=	1 square yard.
272¼ " "	=	1 pole, rod, or perch.
4,356 " "	=	1 square chain.
10,890 " "	=	1 square rood.
43,560 " "	=	1 acre.
27,878,400 " "	=	1 square mile.

No. 4.—Solid Measure.

1728 cubic inches	=	1 cubic foot.
27 cubic feet	=	1 cubic yard.

$16\frac{1}{2}$ feet long, 1 foot high, and $1\frac{1}{2}$ feet thick = 1 perch stone = $24\frac{3}{4}$ cubic feet.

No. 5.—Troy Weight.

Platinum, gold, silver, and some precious stones are weighed by Troy weight, diamonds by carats of 4 grains each.

24 grains	=	1 pennyweight.
20 pennyweights	=	1 ounce (480 grains).
12 ounces	=	1 pound (5760 grains).

No. 6.—Avoirdupois Weight.

16 drams	=	1 ounce (437½ grains).
16 ounces	=	1 pound (7000 grains).
14 pounds	=	1 stone.
2 stones	=	1 quarter.
4 quarters	=	1 hundred-weight (112 pounds).
20 hundred-weight	=	1 ton (long ton) (2240 pounds).

No. 7.—Weight by Specific Gravity.

Frequently the weight of masses is required where it is very inconvenient, or, perhaps, impossible to use scales. The following method may be sufficiently accurate:—

Find the average specific gravity of the mass either by actual weight of a piece or by the following table. Then measure the cubic contents of the mass as nearly as possible and multiply by the weight of a cubic foot. Thus, a mass of limestone (say good marble) measures 40 cubic feet. The specific gravity of good marble is 2.6, that is, it is 2.6 as heavy as a cubic foot of water, which weighs 62.5 pounds. Therefore

$$
\begin{array}{r}
62.5 \\
2.6 \\
\hline
3750 \\
1250 \\
\hline
162.50
\end{array}
$$

A cubic foot of good marble weighs 162.5 pounds, and the 40 cubic feet will weigh

$$
\begin{array}{r}
162.5 \\
40 \\
\hline
6500.0
\end{array}
$$

or about $3\frac{1}{4}$ tons. Of course all rock masses have not plane sides, and the irregularity requires some calculation and various allowances which the prospector must make, and can easily do with a little consideration.

When greater accuracy of specific gravity and of bulk is desired for small masses, and no scales are at hand, the following plan may be very satisfactorily adopted. Fill a tub or hogshead or large box with rain water, after having inserted a tube or piece of tin pipe into the upper edge. Pour in more water until it will hold no more without running out of the spout. Introduce the mass of rock and catch all the water which runs out of the pipe. Now measure the overflow; this represents the exact cubic measure of the rock introduced.

1 gallon contains . . . . 231 cubic inches.
1 quart   "   . . . . 57.75 or $57\frac{3}{4}$ cubic inches.
1 pint   "   . . . . 28.87 or $28\frac{5}{8}$ "   "
1 gill   "   . . . . 7.21 or $7\frac{1}{5}$ "   "

See Appendix, No. 8.

Suppose the overflow was 8 gallons, 1 quart, $4\frac{1}{3}$ gills, and that the specific gravity of the rock or ore was 6.5 by the table below. Then the mass will cause an overflow of 1936.99 cubic inches, and this is 208.99 more than one cubic foot, or about 1.120 of a cubic foot for the mass.

Since 6.5 was the specific gravity of the ore, 6.5×62.5 pounds $= 406.25$, which would be the weight of a cubic foot of the ore, and $406.25 \times 1.120 = 455$ pounds, the exact weight of the mass you introduced into the water.

SPECIFIC GRAVITY, HOW TO FIND. Where the mass is of very nearly the same density in all parts, the specific gravity may be taken of a small part as follows:

Suspend the scales so that they will be steady, weigh about an ounce or pound of the ore accurately, then tie the ore by a horse-hair or a fine silk thread to the hook that holds one of the scales, and let it (the ore) hang below the scale pan, and then weigh the ore entirely submerged in water. The thread or hair may be attached to the centre of the scale pan and weighed in that way, but the pan in either case must remain on the scales just as before. Then the weight in air divided by the weight in air *minus the weight in water*, is the specific gravity; *e. g.*, a piece of ore weighs in air 100 grains, in water 80 grains, then 100 divided by (100—80 = 20) = 5, the specific gravity of that piece of ore. You may now proceed as in the case of the marble block.

No. 8.—SPECIAL WEIGHTS, ETC.

One cubic foot of water is equal to 7.475 U. S. gals. of 231 cubic inches each, or $7\frac{1}{2}$ gallons nearly; or 6.2321 Imperial gals. of $277\frac{1}{4}$ cubic inches each. This, with what follows, is important in the construction of tanks, pools, etc., where contents, weight, and pressure are to be considered.

It should be remembered that, although the English Imperial gallon is $277\frac{1}{4}$ cubic inches = 10 lbs. avoir. of distilled water at 62° Fah., Bar. 30 inches, and equal to 277.274 cubic inches, the United States standard gallon is 231 inches, or 58372.1754 grains, or 8.3389 lbs. of distilled water maximum density. This is almost exactly $=$ to a cylinder 7 inches diameter, 6 inches high. The beer gallon $=$ 282 inches.

One gallon $=$ 8.3389 lbs.; one quart $=$ 2.847 lbs.; one pint $=$ 1.423 lbs.; one gill $=$ .355 lbs.; U. S. standard measure. One cubic foot of water $=$ 62.3210 lbs., British weight; recent and correct, 62.278.

No. 9.—French Measures.—Length.

Millimetre ($\frac{1}{1000}$ of a metre) =		.03937 inch.
Centimetre ($\frac{1}{100}$ " " =		.3937 "
Decimetre ($\frac{1}{10}$ " " =		3.937 "
Metre (the unit of length) =		39.3708 " or 3.2809 ft.
Decametre (10 metres) =		32.809 ft. or 10.9363 yds.
Hectometre (100 metres) =		109.3633 yards.
Kilometre (1000 metres) =		1093.63 yds. or .62138 mile.
Myriametre (10,000 metres) =		6.2138 miles.

Surface.

Centiare ($\frac{1}{100}$ of an are or sq. metre =		1.1960 sq. yds.
Are (unit of surface)	=	{ 119.6033 sq. yards or .0247 acre.
Decare (10 acres)	=	{ 1196.033 sq. yards or .2474 acre.
Hectare (100 ares)	=	{ 11960.33 sq. yards or 2.4736 acres.

Solid Measure.

Decistere ($\frac{1}{10}$ of a stere) =		3.5317 cubic feet.
Stere (cubic metre) =		35.3166 " "
Decastere (10 steres) =		353.1658 " "

WEIGHT.

Milligramme ($\frac{1}{1000}$ of a gramme) = .0154 grain.
Centigramme ($\frac{1}{100}$ "   ) = .1544 grain.
Decigramme ($\frac{1}{10}$ "   ) = 1.544 grains.
Gramme (unit of weight)   = 15.44 grains.
Decagramme (10 grammes)   = 154.4 grains.

Hectogramme (100 "   )   = 1,544 grains. $\begin{cases} 3.2167 \text{ ozs.} \\ \text{Troy or} \\ 3.5291 \text{ ozs.} \\ \text{Avoir.} \end{cases}$

Kilogramme (1000 "   )   = 32¼ ozs. or 2.2057 pounds.
Myriagramme (10,000 grammes) = 22.057 pounds.

No. 10.—SPECIFIC GRAVITY OF METALS, ORES, ROCKS, ETC.

Platinum 16–21
Gold 16–19.5
Mercury 13.5
Lead 11.35–11.5
Silver 10.1–11.1
Copper 8.5–8.9
Iron when pure 7.78
Iron, cast, average . . . . 6.7; foundry 6.9 to 7

ORES : associated with gold and silver.

(Gold) Iron pyrites 4.8–5.2
Copper pyrites 4.0–4.3
(Silver) Galena 7.2–7.7
Glance (silver) 7.2–7.4
Ruby silver (dark) 5.7–5.9
"   "   (light) 5.5–5.6
Brittle silver (sulphide) 5.2–6.3
Horn silver 5.5–5.6

OTHER ORES.

Zinc blende 3.7–4.2
Mercury (Cinnabar) 8.8–9.9
Tin—tinstone, cassiterite 6.4–7.6

Tin pyrites 4.3–4.5
Copper—Red or ruby copper 5.7–6.15
      Gray 5.5–5.8
      Black oxide 5.2–6.3
      Pyrites 4.1–4.3
      Carbonate (Malachite) 3.5–4.1
Lead—sulphide (Galena) 7.2–7.7
      Carbonate (white lead) 6.4–6.6
Zinc—Blende 3.7–4.2
      Calamine 4.0–4.5
Iron—Hematite (red) 4.5–5.3
      Magnetic 4.9–5.9
      Brown hematite 3.6–4.0
      Spathic (carbonate) 3.7–3.9
      Pyrites (mundic) 4.8–5.2
Antimony—gray sulphide 4.5–4.7
Nickel—Kupfer nickel 7.3–7.5
Cobalt—Tin-white 6.5–7.2
      Glance 6.0
      Pyrites 4.8–5
      Bloom 2.91–2.95
      Earthy 3.15–3.29
Manganese—Black oxide 4.7–5.0
      Wad, Bog manganese 2.0–4.6
Bismuth—Sulphide 6.4–6.6
      Oxide **4.3**

MINERALS OF COMMON OCCURRENCE.

Quartz 2.5–2.8
Fluorspar 3.0–3.3
Calc spar 2.5–2.8
Barytes 4.3–4.8
Granite ⎫
Gneiss ⎭ 2.4–2.7
Mica slate 2.6–2.9
Syenite 2.7–3.0
Greenstone trap 2.7–3.0
Basalt 2.6–3.1
Porphyry 2.3–2.7
Talcose slate 2.6–2.8

Clay slate 2.5–2.8
Chloritic slate 2.7–2.8
Serpentine 2.5–2.7
Limestone and Dolomite 2.5–2.9
Sandstones 1.9–2.7
Shale 2.8

Other minerals are mentioned in the text with their specific gravities.

No. 11.—A Ton Weight of the Following will Average in Cubic Feet:

Earth	21 cubic feet.	Pit sand	22 cubic feet.
Clay	18 " "	River sand	19 " "
Chalk	14 " "	Marl	18 " "
Coarse gravel	19 " "	Shingle	23 " "

Power for Mills.

As the Pelton wheel seems to find the most frequent application in California, it may be convenient to have the following rule, applicable to this wheel:

When the head of water is known in feet, multiply it by 0.0024147, and the product is the horsepower obtainable from one miner's inch of water.

The power necessary for different mill parts is:

For each 850 lbs. stamp, dropping 6 inches 95 times per minute . . . . 1.33 H. P.
For each 750 lbs. stamp, dropping 6 inches 95 times per minute . . . . 1.18 "
For each 650 lbs. stamp, dropping 6 inches 95 times per minute . . . . 1.00 "
For an 8-inch by 10-inch Blake pattern rock breaker . . 9.00 "
For a Frue or Triumph vanner with 220 revolutions per minute . . . . 0.50 "
For a 4-foot clean-up pan, making 30 revolutions per minute . . . . 1.50 "

For an amalgamating barrel, making 30 revolutions per
minute . . . . 2.50 "
For a mechanical batea, making 30 revolutions per
minute . . . . 1.00 "

BORING.

Rock is bored with jumpers of 10 to 18 lbs., used
alone or with boring bars and hammer. The
former are more effective, but can only be used
perpendicularly, or nearly so, and with rock of
moderate hardness; they require more skill.

18 lb. hammers are used for 3 inch boring bars.
16 lb.   "   "   " $2\frac{1}{2}$ inch boring bars.
14 lb.   "   "   "   2 and $1\frac{3}{4}$ inch boring bars.
5 to 7 lb.   "   "   "   1 inch boring bars.

The boring bars may be made of $1\frac{1}{8}$-inch bar
iron of various lengths, with steel bits up to 3
inches. A bit should bore from 18 to 24 feet with
each steeling, and requires to be sharpened once for
every foot bored.

DIAMOND DRILL.

This drill is applicable to sinking a bore-hole for
prospecting for minerals or water, shafts, etc., or
blasting under water.

It consists of a circular row of "carbonados," a
species of diamond, set in a circular steel ring.
This is attached to a hollow steel tube, which is
kept rotating at about 250 revolutions per minute,
pressed forward by a force varying from 400 to 800
lbs., according to the nature of the rock. Water is
supplied through the tube, which washes out the
debris and cools the diamonds.

Granite and the hardest limestones are penetrated at the rate of 2 to 3 inches per minute, sandstones 4 inches, quartz 1 inch.

The diamond drill is not effective in soft strata, such as clay, sand and alluvial deposits.

THE CHEMICAL ELEMENTS, THEIR SYMBOLS, EQUIVALENTS AND SPECIFIC GRAVITIES.

Name.	Symbol.	Atomic Weight.	Specific Gravity.
Aluminium	Al.	27.5	2.56
Antimony	Sb.	122.0	6.70
Arsenic	As.	75.0	5.70
Barium	Ba.	137.0	4.00
Bismuth	Bi.	210.0	9.7
Boron	B.	11.0	2.63
Bromine	Br.	80.0	5.54
Cadmium	Cd.	112.0	8.60
Caesium	Cs.	133.0	1.88
Calcium	Ca.	40.0	1.58
Carbon	C.	12.0	3.50
Cerium	Ce.	92.0	6.68
Chlorine	Cl.	35.5	2.45
Chromium	Cr.	52.5	6.81
Cobalt	Co.	58.8	7.7
Columbium	Cb.	184.8	6.00
Copper	Cu.	63.5	8.96
Didymium	Di.	96.0	6.54
Erbium	E.	112.6	—
Fluorine	F.	19.0	1.32
Gallium	Ga.	69.9	5.9
Glucinum	Gl.	9.5	2.1
Gold (Aurum)	Au.	196.7	19.3
Hydrogen	H.	1.0	0.069
Indium	In.	113.4	7.4
Iodine	I.	127.0	4.94
Iridium	Ir.	198.0	21.15
Iron (Ferrum)	Fe.	56.0	7.79
Lanthanum	La.	90.2	11.37
Lead (Plumbum)	Pb.	207.0	11.44

The Chemical Elements, their Symbols, Equivalents and Specific Gravities.

Name.	Symbol.	Atomic Weight.	Specific Gravity.
Lithium	Li.	7.0	0.59
Magnesium	Mg.	24.0	1.75
Manganese	Mn.	55.0	8.01
Mercury (Hydrargyrum)	Hg.	200.0	13.59
Molybdenum	Mb.	96.0	8.60
Nickel	Ni.	58.8	8.60
Niobium	Nb.	94.0	6.27
Nitrogen	N.	14.0	0.972
Osmium	Os.	199.0	21.40
Oxygen	O.	16.0	1.105
Palladinm	Pd.	106.5	11.60
Phosphorus	P.	31.0	1.83
Platinum	Pt.	197.4	21.53
Potassium (Kalium)	K.	39.0	0.865
Rhodium	Ro.	104.3	12.1
Rubidum	Rb.	85.4	1.52
Ruthenium	Ru.	104.4	11.4
Selenium	Se.	79.5	4.78
Silicon	Si.	28.0	2.49
Silver (Argentum)	Ag.	108.0	10.5
Sodium (Natrium)	Na.	23.0	0.972
Strontium	Sr.	87.6	2.54
Sulphur	S.	32.0	2.05
Tantalium	Ta.	182.0	10.78
Tellurium	Te.	129.0	6.02
Thallium	Tl.	204.0	11.91
Thorium	Th.	115.7	7.8
Tin (Stannum)	Sn.	118.0	7.28
Titanium	Ti.	50.0	4.3
Tungsten (Wolfram)	W.	184.0	17.6
Uranium	U.	120.0	18.4
Vanadium	V.	51.3	5.50
Yttrium	Y.	61.7	—
Zinc	Zn.	65.0	7.14
Zirconium	Zr.	89.5	4.15

The figures indicating the proportions by weight in which the elements unite with one another are called the combining or atomic weights, because they represent the relative weights of the atoms of the different elements. Since hydrogen is the lightest element, it is taken as the standard, and its combining or atomic weight = 1.

To find the proportional parts by weight of the elements of any substance whose chemical formula is known:

RULE.—Multiply together the equivalent and the exponent of each element of the compound; the product will be the proportion by weight of that element in the substance.

Example,—Find the proportionate weights of the elements of alcohol, C_2H_6O :

Carbon C_2 = equivalent 12 × exponent 2 = 24
Hydrogen H_6 =   "   1 ×   "   6 = 6
Oxygen O =   "   16 ×   "   1 = 16

Of every 46 lbs. of alcohol, 6 lbs. will be H; 16 O; 24 C.

To find the proportions by *volume*, divide by the specific gravity.

COMMON NAMES OF CHEMICAL SUBSTANCES.

Common Names.	Chemical Names.
Aqua fortis.	Nitric acid.
Aqua regia.	Nitro-hydrochloric acid.
Blue vitriol.	Sulphate of copper.
Cream of tartar.	Bitartrate of potassium.
Calomel.	Chloride of mercury.
Chalk.	Carbonate of calcium.
Caustic potash.	Hydrate of potassium.
Chloroform.	Chloride of formyl.
Common salt.	Chloride of sodium.

Copperas and green vitriol.	Sulphate of iron.
Corrosive sublimate.	Bichloride of mercury.
Dry alum.	Sulphate of aluminium and potassium.
Epsom salts.	Sulphate of magnesium.
Ethiops mineral.	Black sulphide of mercury.
Galena.	Sulphide of lead.
Glauber's salt.	Sulphate of sodium.
Glucose.	Grape sugar.
Iron pyrites.	Bisulphide of iron.
Jeweler's putty.	Oxide of tin.
King's yellow.	Sulphide of arsenic.
Laughing gas.	Protoxide of nitrogen.
Lime.	Oxide of calcium.
Lunar caustic.	Nitrate of silver.
Mosaic gold.	Bisulphide of tin.
Muriate of lime.	Chloride of calcium.
Muriatic acid.	Hydrochloric acid.
Nitre or saltpetre.	Nitrate of potash.
Oil of vitriol.	Sulphuric acid.
Potash.	Oxide of potassium.
Realgar.	Sulphide of arsenic.
Red lead.	Oxide of lead.
Rust of iron.	Oxide of iron.
Sal ammoniac.	Chloride of ammonia.
Salt of tartar.	Carbonate of potassium.
Slaked lime.	Hydrate of calcium.
Soda.	Oxide of sodium.
Spirits of hartshorn.	Ammonia.
Spirits of salt.	Hydrochloric acid.
Stucco or plaster of Paris.	Sulphate of lime.
Sugar of lead.	Acetate of lead.
Verdigris.	Basic acetate of copper.
Vermilion.	Sulphide of mercury.
Vinegar.	Acetic acid (diluted).
Volatile alkali.	Ammonia.
Water.	Oxide of hydrogen.
White precipitate.	Ammoniated mercury.
White vitriol.	Sulphate of zinc.

18

PROSPECTORS' POINTERS.

Take a soft pine board, and a hard lead pencil, and the writing will sometimes outlast your claim. I have seen such notices that have withstood the storms of seven or eight years and still remain legible. There is a great variety of ways to write a notice; and nearly every prospector has his own way. But the briefest and most concise way is as good as any, and the easiest. Now, I'll write you one for the Catharine this way:

CATHARINE LODE.

NOTICE IS HEREBY GIVEN that I, the undersigned citizen of the United States, having complied with Chapter 36, Title 32, Revised Statutes of the United States, and the local regulations of Barker district, claim by right of discovery, 1500 feet in length, and 600 feet in width, along the mineral-bearing vein, to be known as the Catharine (or any other name).

Beginning at centre of discovery shaft and running: "How far do you run northerly?"

"Seven hundred feet northeast."

"Seven hundred feet in a northerly direction and 300 feet in a southerly direction.

"Always say northerly, southerly, easterly, and westerly in writing notices. Don't give it any specific direction. When you say 'northerly,' it gives you a chance to swing your stakes all around the

North Pole, if necessary. You can swing your
stakes after your location is made any way you
want to, provided there are no conflicting claims,
unless you change from northerly and southerly to
easterly and westerly, or vice versa. In that case,
you have to make an amended location and record
it. Let's see. Where were we? Oh, yes; together
with 300 feet on either side of the vein.

"Located this 18th day of June, 1891."

"Locator—TENDERFOOT, Prospector."

"Now that is all that is necessary to hold any
claim, as far as the notice goes. Some prospectors
put in a claim for all dips, spurs, angles, and variations throughout the width, breadth and depth of
the claim; but that's all foolishness. The law
grants you all the spurs and angles and dips you
want. You just go ahead and do as the law requires you to do, to hold any mining claim."—
Butte Bystander.

GLOSSARY OF TERMS

PROSPECTING, MINING, MINERALOGY, GEOLOGY, ETC.

Abraded. Reduced to powder.

Acicular. Needle-shaped.

Adamantine. Of diamond lustre.

Adit. A nearly horizontal passage from the surface by which a mine is entered. In the United States an adit is usually called a *tunnel.*

Aerolite. A stone or other body which has come to the earth from distant space.

Agate. Name given to certain siliceous minerals.

Aggregation. A coherent group.

Alligator. A rock-breaker operating by jaws.

Alloy. A compound of two or more metals fused together.

Alluvium. The earthy deposit made by running streams, especially in times of flood.

Amalgamation. The production of an amalgam or alloy of mercury; also the process in which gold and silver are extracted from pulverized ores by producing an amalgam from which the mercury is afterwards expelled.

Amorphous. Without any crystallization or definite form.

Amygdaloids. Small almond-shaped vesicular cavities in certain igneous rocks, partly or entirely filled with other minerals.

Analysis (in Chemistry). An examination of the substance to find out the nature of the component parts and their quantities. The former is called qualitative and the latter quantitative analysis.

Anemometer. An instrument for measuring the rapidity of an air-current.

Anticlinal. The line of a crest, above or under ground, on the two sides of which the strata dip in opposite directions. The converse of *synclinal.*

Apex. In the U. S. Revised Statutes, the end or edge of a vein nearest the surface.

Aqua fortis. Name formerly applied to nitric acid.

Aqua regia. A mixture of nitric and hydrochloric acids. One volume of strong nitric to three or four of hydrochloric acid is a good mixture.

Arborescent. Of a tree-like form.

Arenaceous. Siliceous or sandy (of rocks).

Argentiferous. Containing silver.

Argillaceous. Containing clay.

Arrastre. Apparatus for grinding and mixing ores by means of a heavy stone dragged around upon a circular bed. Chiefly used for ores containing free gold.

Arsenite. Compound of a metal with arsenic.

Assay. To test ores and minerals by chemical or blowpipe examination.

Assay-ton. A weight of 29.166$\frac{2}{3}$ grammes.

Assessment-work. The work done annually on a mining claim to maintain possessory title.

Auriferous. Containing gold.

Axe Stone. A species of jade. It is a silicate of magnesia and alumina.

Back of a lode. The part between the roof and the surface.

Back-shift. The second set of miners working in any spot each day.

Bank claim. A mining claim on the bank of a stream.

Banket. Auriferous conglomerates cemented together with quartz.

Bar. A vein or dike crossing a lode; also a sand or rock ridge crossing the bed of a stream.

Bar-diggings. Gold-washing claims located on the bars (shallows) of a stream, and worked when the water is low, or otherwise with the aid of coffer-dams.

Barilla. Native copper disseminated in grains in copper ores.

Barrel-amalgamation. The amalgamation of silver ores in wooden barrels with quicksilver, metallic iron, and water.

Base metals. The metals not classed as *noble* or precious. See *Noble metals.*

Bases. Compounds which are converted into salts by the action of acids.

Basin. A natural depression of strata containing a coal bed or other stratified deposit; also the deposit itself.

Battery. A set of *stamps* in a stamp mill comprising the number which fall in one *mortar*, usually five; also a bulkhead of timber.

Battery-amalgamation. Amalgamation by means of mercury placed in the mortar.

Bed. A seam or deposit of mineral, later in origin than the rock below, and older than the rock above; that is to say, a regular member of the series of formation, and not an intrusion.

Bedded-vein. A lode occupying the position of a bed, that is, parallel with the stratification of the inclosing rocks.

Bed-rock. The solid rock underlying alluvial and other surface formations.

Bed-way. An appearance of stratification, or parallel marking, in granite.

Belly. A swelling mass of ore in a lode.

Black band. A variety of carbonate of iron.

Black flux. A mixture of charcoal and potassium carbonate.

Black jack. Zinc-blende.

Black tin. Tin ore ready dressed for smelting.

Blanch. Lead ore mixed with other minerals.

Blanched copper. An alloy of copper and arsenic.

Blende. Sulphide of zinc.

Blind level. A level not yet connected with other workings.

Blind lode. One that does not show surface croppings.

Blossom. The oxidized or decomposed outcrop of a vein or coal bed. Also called *smut* and *tailing.*

Blow-out. 1. A large outcrop beneath which the vein is smaller. 2. A shot or blast is said to blow out when it goes off like a gun, and does not shatter the rock.

Blue-john. Fluorspar.

Blue lead. The bluish auriferous gravel and cement deposit found in the ancient river-channels of California.

Bluff. A high bank or hill with a precipitous front.

Bonanza. A body of rich ore.

Booming. The accumulation and sudden discharge of a quantity of water (in placer mining, where water is scarce). See also *Hushing.*

Bort. Opaque black diamond.

Botryoidal. Like a bunch of grapes.

Boulder. A fragment of rock brought by natural means from a distance, and usually large and rounded in shape.

Brasque. A lining for crucibles; generally a compound of clay, etc., with charcoal dust.

Breast. The face of a working.

Breccia. A *conglomerate* in which the fragments are angular.

Buddle. An inclined vat, or stationary or revolving platform upon which ore is concentrated by means of running water.

Bullion. Uncoined gold and silver. *Base bullion* is pig lead containing silver and some gold, which are separated by refining.

Buried rivers. River beds which have been buried below streams of basalt or alluvial drifts.

Burr. Solid rock.

Button. The globule of metal remaining in a crucible at the end of fusion.

Cage. A frame with one or more platforms used in hoisting in a vertical shaft.

Cairngorm. A variety of quartz, frequently transparent; used as an ornament.

Calcareous. Containing carbonate of lime.

Calcination. Roasting at a gentle heat.

Calcine. To expose to heat with or without oxidation.

Calcite. **Carbonate of lime.**

Cañon. A valley, usually precipitous; a gorge.

Cap or cap-rock. Barren vein matter, or a *pinch* in a vein, supposed to overlie ore.

Carat. Weight, nearly equal to four grains, used for diamonds and precious stones. With goldsmiths and assayers the term carat is applied to the proportions of gold in an alloy; 24 carats represents fine gold. Thus 18 carat gold signifies that 18 out of 24 parts are pure gold, the rest some other metal.

Carbonaceous. Containing carbon not oxidized.

Carbonates. The common term in the West for ores containing a considerable proportion of carbonate of lead.

Carbonization. Conversion to carbon.

Case. A small fissure admitting water into the workings.

Casing. Clayey material found between a vein and its wall.

Cawk. Sulphate of baryta (heavy spar).

Cement. Gravel firmly held in a siliceous matrix, or the matrix itself.

Champion lode. The main vein as distinguished from branches.

Chasing. Following a vein by its range or direction.

Chert. Hornstone; a siliceous stone often found in limestone.

Choke damp. Carbonic acid gas.

Chlorides. A common term for ores containing chloride of silver.

Chloridize. To convert into chloride. Applied to the roasting of silver ores with salt, preparatory to amalgamation.

Chute. A channel or shaft underground, or an inclined trough above ground, through which ore falls or is "shot" by gravity from a lower to a higher level.

Claim. The portion of mining ground held under the Federal and local laws by one claimant or association, by virtue of one location and record.

Clay slate. A slate formed by the induration of clay.

Cleavage. The property of a mineral of splitting more easily in some directions than in others.

Cleavage planes. The planes along which cleavage takes place.

Clinometer. An apparatus for measuring vertical angles, particularly *dips.*

Cobre ores. Copper ores from Cuba.

Color. A particle of gold found in the prospector's pan.

Concentration. The removal by mechanical means of the lighter and less valuable portions of ore.

Conchoidal. Name given to a certain kind of fracture resembling a bivalve shell.

Concretion. A nodule formed by the aggregation of mineral matter from without round some centre.

Conglomerate. A rock consisting of fragments of other rocks (usually rounded) cemented together.

Consume. The chemical and mechanical loss of mercury in amalgamation.

Contact. The plane between two adjacent bodies of dissimilar rock. A *contact-vein* is a vein, and a *contact-bed* is a bed, lying, the former more or less closely, the latter absolutely, along a contact.

Contortion. Crumpling and twisting.

Coprolites. Phosphate of lime; petrified excrements of animals.

Counter. A cross vein.

Country, or *Country rock.* The rock traversed by or adjacent to an ore deposit.

Course of a lode. Its direction.

Cradle. See *Rocker*.

Cranch. Part of a vein left by old workers.

Crate dam. A dam built of crates filled with stone.

Crater. The cup-like cavity at the summit of a volcano.

Cretaceous. Chalky.

Crevet. A crucible.

Crevice. A shallow fissure in the bed-rock under a gold placer, in which small but highly concentrated deposits of gold are found; also the fissure containing a vein.

Cribbing. Close timbering, as the lining of a shaft.

Cribble. A sieve.

Cropping-out. The rising of layers of rock to the surface.

Cross-course. An intersecting (usually), a barren vein.

Cross-cut. A level driven across the course of a vein.

Cross-vein.   An intersecting vein.

Cupriferous.   Containing copper.

Cyanidation.   Conversion of gold into a double cyanide of potassium and gold by the action of cyanide of potassium.

Dead-roasting.   Roasting carried to the farthest practicable degree in the expulsion of sulphur.

Dead-work.   Work that is not directly productive, though it may be necessary for exploration and future production.

Debris.   The fragments resulting from shattering and disintegration.

Decrepitate.   To crackle and fly to pieces when heated.

Deep Leads.   Alluvial deposits of gold or tinstone buried below a considerable thickness of soil or rock.

Delta.   The alluvial land at the mouth of a river; usually bounded by two branches of the river, so as to be of a more or less triangular form.

Dentritic.   Like branches of trees.

Denudation.   Rock laid bare by water or other agency.

Deoxidation.   The removal of oxygen.

Desilverization.   The process of separating silver from its alloys.

Desulphurization.   The removal of sulphur from sulphuret ores.

Detritus.   Accumulations from the disintegration of exposed rock surfaces.

Development.   Work done in opening of a mine.

Dialling.   Surveying a mine by means of a dial.

Diggings.   Applicable to all mineral deposits and mining camps, but in usage in the United States applied to placer-mining only.

Dike.   A vein of igneous rock.

Diluvium.   Sand, gravel, clay, etc., in superficial deposits.

Dip.   The inclination of a vein or stratum below the horizontal.

Disintegration.   The breaking asunder of solid matter due to chemical or physical forces.

Dislocation.   The displacement of rocks on either side of a crack.

Divining rod.   A rod, most frequently of witch-hazel, and forked in shape, used according to an old but still extant superstition for discovering mineral veins and springs of water, and even for locating oil wells.

Discovery. The first finding of the mineral deposit in place upon a mining claim. A *discovery* is necessary before the location can be held by a valid title. The opening in which it is made is called *discovery-shaft*, *discovery-tunnel*, etc.

Ditch. An artificial water-course, flume or canal to convey water for mining.

Dolly. An apparatus used in washing gold-bearing rocks (Australia).

Domes. Strata which are dipping away in every direction.

Drift. A horizontal passage underground ; also *unstratified diluvium.*

Druse. A crystallized crust lining the sides of a cavity.

Dry Ores. Silver ores which do not contain lead.

Dyke. See *Dike.*

Efflorescence. An incrustation of powder or threads, due to the loss of the water of crystallization.

Elements. Substances which have never been decomposed.

Elutriation. Purification by washing and pouring off the lighter matter suspended in water, leaving the heavier portions behind.

Entry. An adit.

Erosion. The act or operation of wearing away.

Excrescence. Grown out of.

Exfoliate. To peel off in leaves from the outside.

Exploitation. The productive working of a mine as distinguished from exploration.

Face. In any adit, tunnel, or slope, the end at which work is progressing or was last done.

False Bottom. In alluvial mining a stratum on which auriferous beds lie, but which has other bottoms below it.

Fathom. 6 feet.

Fault. A dislocation of the strata or vein.

Feather Ore. A sulphide of lead and antimony.

Feeder.   A small vein adjoining a larger vein.

Feldspathic.   Containing feldspar as the principal ingredient.

Ferruginous.   Containing iron.

Fire-damp.   Light carburetted hydrogen gas.

Fissure-vein.   A fissure in the earth's crust filled with mineral.

Flexible.   Capable of being bent without elasticity.

Flint.   A masssive impure variety of silica.

Float-copper.   Fine scales of metallic copper which do not readily settle in water.

Float-gold.   Fine particles of gold which do not readily settle in water, and hence are liable to be lost in the ordinary stamp-mill process.

Float-ore.   Water-worn particles of ore; particles of vein-material found on the surface, away from the vein-outcrop.

Flocculent.   Cloudy, resembling lumps of wool.

Floor.   The rock underlying a stratified or nearly horizontal deposit, also a horizontal flat ore body.

Flume.   A wooden conduit bringing water to a mine or mill.

Flux.   A salt or other mineral added in smelting to assist fusion by forming more fusible compounds.

Foliated.   Arranged in leaf-like laminæ (such as mica schist).

Foot-wall.   The wall under the vein.

Forfeiture.   The loss of possessory title to a mine by failure to comply with the laws prescribing the quantity of *assessment* work, or by actual abandonment.

Formation.   The series of rocks belonging to an age, period or epoch, as the Silurian *formation.*

Fossil.   Term applied to express the animal or vegetable remains found in rocks.

Foundershaft.   The first shaft sunk.

Free.   Native, uncombined with other substances, as free gold or silver.

Free-milling.   Applied to ores which contain free gold or silver, and can be reduced by crushing and amalgamation, without roasting or other chemical treatment.

Fritting. The formation of a slag by heat with but incipient fusion.

Fuller's earth. An unctuous clay, usually of a greenish-gray tint, compact yet friable. Used by fullers to absorb moisture.

Fuse. In blasting the fire is conveyed to the blasting agent by means of a prepared tape or cord called the fuse.

Gad. A steel wedge.

Galiage. Royalty.

Gallery. A level or drift.

Gangue. The mineral associated with the ore in a vein.

Gash. Applied to a vein wide above, narrow below, and terminating in depth within the formation it traverses.

Geode. A cavity, studded around with crystals or mineral matter, or a rounded stone containing such cavity.

Geysers. Intermittent boiling springs.

Glacier. A body of ice which descends from the high to the low ground.

Glance. Literally, shining. Name applied to certain sulphides.

Globule. A small substance of a spherical shape.

Goaves. Old workings.

Gopher or *Gopher-drift.* An irregular prospecting drift, following or seeking the ore without regard to maintenance of a regular grade or section.

Gossan or *Gozzan.* Hydrated oxide of iron, usually found at the decomposed outcrop of a mineral vein.

Gravel mine. In the United States, an accumulation of auriferous gravel.

Grip. A small narrow cavity.

Grit. A variety of sandstone of coarse texture.

Gubbin. A kind of iron stone.

Gulch. A ravine.

Gullet. An opening in the strata.

Hade. See *Underlay.*

Hanging-side or *Hanging-wall,* or *Hanger.* The wall or side over the vein.

Hard Head. A residual alloy containing much iron and arsenic, produced in the refining of tin.

Heading. The vein above a drift; also an interior level or air-way driven in the mine.

Heading side. The under side of a lode.

Heave. An apparent lateral displacement of a lode produced by a fault.

Hog back. A sharp anticlinal, decreasing in height at both ends until it runs out; also a ridge produced by highly tilted strata.

Homogeneous. Of the same structure throughout.

Horse. A mass of *country rock* enclosed in an ore deposit.

Hungry. A term applied to hard barren vein matter, such as white quartz.

Hushing. The discovery of veins by the accumulation and sudden discharge of water, which washes away the surface soil and lays bare the rock. See *Booming.*

Hydraulicking. Washing down a bank of earth or gravel by the use of pipes, conveying water under high pressure.

Hydrous. Containing water in its composition.

Igneous. Resulting from the action of fire, as, lavas and basalt are igneous rocks.

Impregnation. An ore-deposit consisting of the country-rock impregnated with ore.

Incline. A shaft not vertical ; also a *plane*, not necessarily under ground.

Incrustation. A coating of matter.

Indicator Vein. A vein which is not metalliferous itself, but, if followed, leads to ore deposits.

In place. Of rock, occupying, relative to surrounding masses, the position that it had when formed.

In situ. In place where formed.

Intrusion. Forcing through.

Irestone. Hard clay slate: hornstone; hornl-bende.

Iridescent. Showing rainbow colors.

Jigging. Separating ores according to specific gravity with a sieve agitated up and down in water. The apparatus is called a *jig* or *jigger*.

Jinny-road. A gravity plane undergronnd.

Jump. To take possession of a mining claim alleged to have been forfeited or abandoned; also, a dislocation of a vein.

Keckle-meckle. The poorest kind of lead ore.

Kibbal or *kibble.* An iron bucket for raising ore.

Kicker. Ground left in first cutting a vein, for support of its sides.

King's yellow. Sulphide of arsenic.

Knits or *knots.* Small particles of ore.

Lagoon. A marsh, shallow pond or lake.

Lamellar. In thin sheets.

Lamina. A thin plate or scale.

Lava. Rock formed by the consolidation of liquid matter which has flowed from a volcano.

Leaching. See *Lixiviation.*

Leads. The auriferous portions of alluvial deposits marking the former courses of streams.

Leath. Applied to the soft part of a vein.

Lenticular. Lens-like.

Level. A horizontal passage or drift into or in a mine.

Limp. An instrument for striking the refuse from the sieve in washing ores.

Litharge. Protoxide of lead.

Lixiviation. The separation of a soluble from an insoluble material by means of washing with a solvent.

Loadstone. An iron ore consisting of protoxide and peroxide of iron; Magnetite.

Locate. To establish a right to a mining claim.

Lode. A regular vein carrying metal.

Long Tom. A kind of gold-washing cradle.

Magma. Paste or groundwork of igneous rocks.

Mainway.   A *gangway* or principal passage.

Marl.   Clay containing carbonate of lime.

Mass-copper.   Native copper occurring in large masses.

Massicot.   See *Litharge.*

Matrix.   The rock or earthy material containing a mineral or metallic ore; the *gangue.*

Matt or *Matte.*   A mass consisting chiefly of metallic sulphides got in the fusion of ores.

Measures.   Strata of coal, or the formation containing coal beds.

Meat-earth.   The vegetable mould.

Metalliferous.   Metal-bearing.

Metamorphic.   Changed in form and structure.

Mine.   In general, any excavation for minerals.   More strictly, subterranean workings, as distinguished from *quarries, placer* and *hydraulic* mines, and surface or open works.

Mineral.   In miners' parlance, ore.

Mineralized.   Charged or impregnated with metalliferous mineral.

Mineral-right.   The ownership of the minerals under a given surface, with the right to enter thereon, mine and remove them.   It may be separated from the surface ownership, but, if not so separated by distinct conveyance, the latter includes it.

Mine-rent.   The rent or royalty paid to the owner of a mineral right by the operator of the mine.

Miners' inch.   A local unit for the measurement of water supplied to hydraulic miners.   It is the amount of water flowing under a certain head through one square inch of the total section of a certain opening for a certain number of hours daily.

Minium.   Protosesquioxide of lead.

Mock ore.   A false kind of mineral.

Monkey drift.   A small prospecting drift.

Monoclinal.   Applied to any limited portion of the earth's crust throughout which the strata dip in the same direction.

Mountain blue.   Blue copper ore.

Muffle.   A semi-cylindrical or long-arched oven, usually small and made of fire clay.

Mundic. Iron pyrites, called so in Cornwall. White mundic is mispickel.

Nacreous. Resembling mother-of-pearl.

Native. Occurring in nature; not artificially formed; usually applied to the metals.

Nickeliferous or *Niccoliferous.* Containing nickel.

Nittings. The refuse of good ore.

Noble metals. The metals which have so little affinity for oxygen that their oxides are reduced by the mere application of heat without a reagent; in other words, the metals least liable to oxidation under ordinary conditions. The list includes gold, silver, mercury, and the platinum group.

Nodule or *Noddle.* A small round mass.

Nugget. A lump of native metal, especially of a precious metal.

Nucleus. A body about which anything is collected.

Open cut. A surface working, open to daylight.

Ore. A natural mineral compound, of the elements of which one at least is a metal.

Organic Compounds. Compounds containing carbon, generally derived from animals or plants.

Outcrop. The portion of a vein or stratum emerging at the surface, or appearing immediately under the soil and surface dèbris.

Output. The product of a mine.

Oxidation. A chemical union with oxygen.

Oxide. The combination of a metal with oxygen.

Pack Walls. Walls built of loose material in mines to support the roof.

Panning. Washing earth or crushed rock in a pan, by agitation with water, to obtain the particles of greatest specific gravity it contains; chiefly practiced for gold, also for quicksilver, diamonds and other gems.

Parting. The separation of two metals in an alloy, especially the separation of gold and silver by means of nitric or sulphuric acid.

Pavement. The *floor* of a mine.

19

Pay-streak. The zone in a vein which carries the profitable or *pay ore.*

Peroxide. An oxide containing more oxygen than some other oxide of the same element.

Peter or *peter-out.* To fail gradually in size or quality.

Petrified. Changed to stone.

Petrology. Study of rocks.

Phosphates. Phosphoric acid combinations.

Pinch. To contract in width.

Pipe or *pipe-vein.* An ore-body of elongated form.

Piping. Washing gold deposits by means of a hose.

Placer. A deposit of valuable mineral, found in particles in *alluvium* or *diluvium,* or beds of streams, etc.

Plasma. A green variety of quartz.

Plastic. Easily moulded.

Plat. The map of a survey in horizontal projection.

Plumbago. Graphite or black lead.

Plumb Bob. A weight suspended by a string to determine vertical lines.

Plush Copper. A fibrous red copper ore.

Pocket. A small body of ore.

Porphyritic. Of the nature of porphyry.

Potstone. Compact steatite.

Precipitate. Term applied to solid matter which is separated from a solution by the addition of reagents or exposure to heat.

Prill. A good sized piece of pure ore.

Prisms. Solids whose bases are plane figures, and whose sides are parallelograms.

Pryan. Ore in small pebbles mixed with clay.

Pudding-Stone. A conglomerate in which the pebbles are rounded.

Pulp-assay. The assay of samples taken from the *pulp, i. e.,* pulverized ore and water, after or during crushing.

Putty powder. Crude oxide of tin.

Quarry. An open or day working.

Quartz. Crystalline silica ; also, any hard gold or silver ore, as distinguished from gravel or earth, hence *quartz-mining* as distinguished from hydraulic mining, etc.

Quartzose. Containing quartz as a principal ingredient.

Quicksand. Sand which is, or becomes, upon the access of water, "quick," *i. e.*, shifting, easily movable or semi-liquid.

Race. A small thread of spar or ore.

Radiating. Diverging from a centre.

Range. A mineral-bearing belt of rocks.

Ravine. A deep narrow valley.

Reduce. To deprive of oxygen ; also, in general, to treat metallurgically for the production of metal.

Refractory. Resisting the action of heat and chemical agents.

Reniform. Kidney-like.

Reticulated Veins. Veins traversing rocks in all directions.

Reverse Faults. Faults due to thrust; the hanging-wall side of the fault being forced upwards on the foot-wall.

Rider. See *Horse.*

Riffle. A groove or interstice, or a cleat or block, so placed as to produce the same effect, in the bottom of a sluice, to catch free gold.

Rim-rock. The bed-rock rising to form the boundary of a placer or gravel deposit.

Rise. That portion of a bed or coal-seam which lies above a level is said to be "to the rise."

Roasting. Calcination, usually with oxidation.

Rocker. A short trough in which auriferous sands are agitated by oscillation, in water, to collect their gold.

Rolley-way. A *gangway.*

Roof. The strata immediately above a coal seam.

Rosette copper. Disks of copper, red from the presence of sub-oxide, formed by cooling the surface of melted copper through sprinkling with water.

Royalty. The dues of a lessor or landlord of a mine, or of the owner of a patented invention.

Rusty gold. Free gold which does not easily amalgamate, the particles being coated, as is supposed, with oxide of iron.

Saccharoid. Like lump-sugar.

Saddle. An anticlinal in a bed or flat vein.

Sal ammoniac. Chloride of ammonium.

Saline. A salt-spring or well; salt works.

Sampling. Mixing ores so that a portion taken from the mixture may fairly represent the whole body.

Schist. Crystalline rock.

Schorl. Black tourmaline.

Seam. A stratum or bed of coal or other mineral.

Sectile. Easily cut.

Sediment. A deposit formed by water.

Segregate. To separate the undivided joint ownership of a mining claim into smaller individual "segregated" claims.

Segregation. A mineral deposit formed by concentration from the adjacent rock.

Salvage or *Selfedge.* A layer of clay or decomposed rock along a vein-wall.

Shaft. A pit sunk from the surface.

Shake. A cavern, usually in limestone; also a crack in a block of stone.

Shale. Consolidated clay.

Shift. The time for a miner's work in one day; also the gang of men working for that period, as the *day-shift*, the *night-shift*.

Shingle. Clean gravel.

Side-basset. A transverse direction to the line of dip in strata.

Silicates. Compounds of silica or silicic acid with a base.

Siliceous. Consisting of or containing silex or quartz.

Sinter. A deposit from hot springs.

Slag. The vitreous mass separated from the fused metals in smelting ores.

Slate. Indurated clays, sometimes metamorphosed.

Slickensides. Polished and sometimes striated surfaces on the walls of a vein, or on interior joints of the vein-material or of rock masses.

Slide. A fault or cross course.

Slime ore. Finely crushed ore mixed with water to the consistence of mud or slime.

Sline. Natural transverse cleavage of rock.

Slip. A vertical dislocation of rocks.

Slope. An inclined opening to a mine.

Sluicing. Washing auriferous earth through long boxes (*sluices*).

Slums. The most finely crushed ores.

Spall or *Spawl.* To break ore. Pieces of ore thus broken are called *spalls.*

Speiss or *speise.* Impure metallic arsenides, principally of iron, produced in copper and lead smelting. Cobalt and nickel are found concentrated in the speiss obtained from ores containing these metals.

Spoon. An instrument made of an ox or buffalo horn, in which earth or *pulp* may be delicately tested by washing to detect gold, amalgam, etc.

Spur. A branch leaving a vein, but not returning to it.

Stalactites. Icicle-like incrustations hanging down from the roof of caves.

Stalagmites. Similar to stalactites, but formed on the floor of the caves by the deposition of solid matter held in solution by dropping water.

Stannary. A tin mine, or tin works.

Step-vein. A vein alternately cutting through the strata of country-rock and running parallel with them.

Stockwork. An ore deposit of such a form that it is worked in floors or stories.

Stope. To remove the ore.

Stratum. A bed or layer.

Streak. The powder of a mineral, or the mark which it makes when rubbed upon a harder substance.

Striated. Marked with parallel grooves or *striæ.*

Strike. The direction of a horizontal line drawn in the middle plane of a vein or stratum not horizontal.

String. A small vein.

Strip. To remove from a quarry, or open working, the overlying earth and disintegrated or barren surface rock.

Stull. A platform laid on timbers, braced across a working from side to side, to support workmen or to carry ore or waste.

Sturt. A *tribute*-bargain which turns out profitable for the miner.

Sublimation. The volatilization and condensation of a solid substance without fusion.

Submetallic. Of imperfect metallic lustre.

Subsidence. The sinking down of.

Subtransparent. Of imperfect transparency.

Sulphate. A salt containing sulphuric acid.

Sulphide. A combination of metal with sulphur.

Sulphurets. In miners' phrase, the undecomposed metallic ores, usually sulphides. Chiefly applied to auriferous pyrites.

Synclinal. The axis of a depression of the strata; also the depression itself. Opposed to *anticlinal*, which is the axis of an elevation.

Tailings. The lighter and sandy portions of the ore on a buddle or in a sluice.

Tail-race. The channel in which tailings, suspended in water, are conducted away.

Thermal. Hot, *e. g.*, thermal springs.

Throw. A dislocation or fault of a vein or stratum, which has been *thrown up or down* by the movement.

Tinstone. Ore containing small grains of oxide of tin.

Toadstone. A kind of trap-rock.

Toughening. Refining, as of copper or gold.

Translucent. Allowing light to pass through, yet not transparent.

Trap. In miners' parlance, any dark igneous, or apparently igneous, or volcanic rock.

Trend. The course of a vein.

Tribute. A portion of ore given to the miner for his labor.

Trogue. A wooden trough, forming a drain.

Trow. A wooden channel for air or water.

Tuff or *Tufa.* A soft sandstone or calcareous deposit.

Tunnel. A nearly horizontal underground passage, open at both ends to day. See *Adit.*

Turn. A pit sunk in a *Drift.*

Underlay or *Underlie.* The departure of a vein or stratum from the vertical, usually measured in horizontal feet per fathom of inclined depth.

Unstratified. Not arranged in strata.

Upcast. The lifting of a coal seam by a dike.

Vanning. Washing " tin-stuff" by means of a shovel.

Vein. See *Lode.* The term *vein* is also sometimes applied to small threads, or subordinate features of a larger deposit.

Vein stuff. Ore associated with gangue.

Vermilion. Mercury sulphide.

Vitreous. Glassy.

Volatile. Capable of easily passing off as vapor.

Vug, Vugg or *Vugh.* A cavity in the rock, usually lined with a crystalline incrustation. See *Geode.*

Walls. The boundaries of a lode, the upper one being the " hanging," the lower the " foot wall."

Wash Dirt. Auriferous gravel, sand, clay, etc.

Wastrel. A tract of waste land, or any waste material.

Weathering. Changing under the effect of continued exposure to atmospheric agencies.

Whim or *Whimsey.* A machine for hoisting by means of a vertical drum, revolved by horse or steam power.

White-damp. A poisonous gas sometimes encountered in coal mines.

Wild lead. Zinc blende.

Win. To extract ore or coal.

Wing Dams. Dams built from the side of a river with the object of deflecting it from its course.

Winze. An interior shaft, usually connecting two levels.

Working home. Working toward the main shaft in extracting ore.

Working out.   Working away from the main shaft in extracting ore.

Zinc-scum.   The zinc-silver alloy skimmed from the surface of the bath in the process of desilverization of lead by zinc.

Zinc-white.   Oxide of zinc.

INDEX.

20